Imperium Press was founded in 2018 to supply students and laymen with works in the history of rightist thought. If these works are available at all in modern editions, they are rarely ever available in editions that place them where they belong: outside the liberal weltanschauung. Imperium Press' mission is to provide right thinkers with authoritative editions of the works that make up their own canon. These editions include introductions and commentary which place these canonical works squarely within the context of tradition, reaction, and counter-Enlightenment thought—the only context in which they can be properly understood.

BREEDING THE HUMAN HERD

EUGENICS, DYSGENICS AND THE FUTURE OF THE SPECIES

EDWARD DUTTON

PERTH
IMPERIUM PRESS
2023

Published by Imperium Press
www.imperiumpress.org

Cover design by Swifty

First Edition

A catalogue record for this
book is available from the
National Library of Australia

ISBN 978-1-922602-84-8 Paperback
ISBN 978-1-922602-85-5 Hardcover
ISBN 978-1-922602-86-2 E-book

In memory of the Rev'd Rick Conrad (1966-2021) and Nicola Ratcliffe (1981-2000).

We civilised men, on the other hand, do our utmost to check the process of elimination; we build asylums for the imbecile, the maimed, and the sick; we institute poor-laws; and our medical men exert their utmost skill to save the life of every one to the last moment. There is reason to believe that vaccination has preserved thousands, who from a weak constitution would formerly have succumbed to small-pox. Thus the weak members of civilised societies propagate their kind. No one who has attended to the breeding of domestic animals will doubt that this must be highly injurious to the race of man. It is surprising how soon a want of care, or care wrongly directed, leads to the degeneration of a domestic race; but excepting in the case of man himself, hardly any one is so ignorant as to allow his worst animals to breed.

—Charles Darwin
The Descent of Man, 1871

Contents

Acknowledgements xv

Chapter One: Allergic to the Basics of Life

The Case of Nicky Ratcliffe 3
The Collapse of Darwinian Selection 5
The Rise of Allergies 8
The Rise of Poor Genetic Health 15

Chapter Two: What Was Being Selected for Under Pre-Industrial Conditions and How?

Jaywick, Essex 21
What is Pleiotropy? 24
Natural Selection 25
How Do Mutations Arise? 27
Epistastic Effects: Bipolar Disorder 28
Recessive: Sickle Cell Anemia 29
Dominant: Huntingdon's Disease 30
X-Linked: Hemophilia and Turner Syndrome 30
Paternal Age and Mutation 32
Maternal Age and Down's Syndrome 33
Sexual Selection 37
Passing on Your Genes Indirectly 39
Refuting Criticisms of Group Selection 43
How Does It All Come Together? 45
The People of 1800 48

Chapter Three: The Rise of Intelligence

The Mechanized Spinning Frame 49
What is Intelligence? 50
What are IQ Tests? 52
Criticisms of the Validity of IQ Tests 53
The Heritability of IQ 55
Selection for Intelligence Under Pre-Modern Conditions: Hunter-Gatherers 56
Pastoralists 59

Early Nation States 60
Jewish Eugenics 61
Hinduism and the Caste System 63
The Christian World 67
Fertility in the Early Modern Era 68
Upward Social Mobility 71
Illegitimates 73
Execution 74
Is there Evidence that Intelligence Was Increasing Up to the Industrial Revolution? 76
Interest Rates 76
Literacy and Numeracy 77
The Size of the Head 79
Per Capita Major Innovation 80

Chapter Four: The Fall of Intelligence
The Collapse of Intelligence in Miami 81
Why Did Intelligence Go Into Decline? 83
Contraception 85
Feminism 86
Welfare 87
IQ and Fertility 88
Evidence for Intelligence Decline and the Flynn Effect 89
Simple Reaction Times 91
Colour Discrimination 93
Use of High-Difficulty Words 94
Backward Digit Span 96
Levels of Genius 97
A Decline in the Frequencies of Variants in the Genome Associated with *g* 99
What About Pollution? 100
The Decline of National IQs and Immigration 101
Eating Into the Capital 102

Chapter Five: The Rise of Pro-Social Personality, Mental Health, and Religiosity
Risking Your Life for Your Country 105
Selection for Pro-Social Personality and for Mental Health 106
The General Factor of Personality 109
How Personality Develops and Why there are Differences 110
Life History Strategy and Personality 113

Ecological Harshness and Selection for Modal Personality 116
Niches and Personality 118
Selection for Religiousness 119
How Did Belief in Gods Develop? 121
Religion as a Selection Pressure 123

Chapter Six: The Fall of Pro-Social Personality, Mental Health, and Religiosity
Screaming Like a Wounded Animal 125
The Breakdown of Selection for Pro-Social Personality 127
Liberalism and Anti-Social Personality Traits 128
Liberalism and Moral Foundations 130
The Break-up of the Religious Bundle 133
The Rise of Genetic Mental Disorders 137
Negative Social Epistasis 141
Evolutionary Mismatch and Dysphoria 144
Why Does Intelligence Predict Not Wanting Children in Advanced Societies? 146
Kinship and Fertility 148
Sexuality and Sexual Orientation 150
Gender Dysphoria 152
Sexual Fantasy and the Self 155
Increasing LGBT Identification 158
Suppressing Atheism 161

Chapter Seven: The Rise and Fall of Genetic Physical Health
A Monstrous Birth 163
The Rise of Developmental Instability 165
The Rise of Cancer and Other Partly Genetic Disorders 168
The Rise of Infertility 172
The Survival of the Least Fit 174
Pre-Natal Screening and "Wrongful Life" 176
The Cost of Genetic Diseases 178

Chapter Eight: The History and Rise of Eugenics
The Runt of the Litter 181
Hunter-Gatherers, Pastoralists and Primitive Farmers 182
Eugenics and the City 185
Plato and Eugenics 187

Slavery in the United States 188
Modern Advocacy for Eugenics: Benedict Morel 190
Francis Galton 191
Eugenics as an Academic Discipline 193
Eugenics Promoted to the Public 194
Eugenics and the Left 197
Family Planning 198
Social Work 199
The Literary World 199
Leftist Politicians 201
Liberals and Eugenics 202
Eugenics, Traditionalism, and the Radical Right 204
Academic Eugenics and Eugenic Laws 207
The United Kingdom 207
The United States 211
Canada, Australia, and South Africa 215
Eugenics and Race in the US and the UK 216
Why Was the Nineteenth Century So Conservative? 218
Eugenics in Continental Europe: Switzerland and Germany 221
Nordic Countries 222
Catholic Europe and Latin America 224
Outside Europe 226
India 226
China 227
Singapore 227
North Korea 228

Chapter Nine: The Fall of Eugenics
Welfare Queens 231
The Wall Street Crash 232
The 1960s: A Shift in Values 233
"Not Even Trying" 236
Feminism and the Suppression of Genius 239
Casting a Large Net 242
The Diehard Eugenicists 245
Mankind Quarterly 246
Richard Lynn 247
Helmuth Nyborg 249
William Shockley 249
Raymond Cattell 250
David Lykken 251

Eugenicist Bioethicists 252
Public Outcries 253
The London Conference on Intelligence 254
Black Lives Matter and COVID-19 257

Chapter Ten: The Future of Eugenics and Dysgenics
The End of Hospitals 259
The Underclass and Medical Chaos 259
Cycles of Civilization 262
Mini-Cycles 265
Population Collapse 266
Slow Life History Strategy 267
Contraception and Abortion 268
Voluntary Euthanasia and Imprisonment 273
In Vitro Fertilization 275
Wokeness 276
The Future of Eugenics 277
The Problems with Eugenics 279
Re-imposing High Child Mortality 281
Would It Even Work? 283
Alternatives to Classical Eugenics 285
Fermi's Paradox 287
Would Civilization Really Collapse? 288
"The Elites Won't Let a Collapse Happen." 289
The Grip of Individualism 291
We Are All Eugenicists 293
Advanced Apes 295
The Great Justinian Plague 297

About the Author 301
Index 303

Acknowledgements

I would like to extend my thanks to J.O.A. Rayner-Hilles, Esq., Prof. Gerhard Meisenberg and Prof. Richard Lynn for reading through earlier versions of this manuscript and offering many helpful comments. My writing of this book has also benefitted from many fruitful discussions especially with Mr George Francis, Herr Emil Kirkegaard, Herr Jakub Kuzminski, Mr Lewis Low, Dr Neema Parvini, Dr David Skrbina, Dr Bo Winegard and Dr Michael Woodley of Menie. Some parts of this book have previously appeared, either in an abridged or in an extended form, in *At Our Wits' End: Why We're Becoming Less Intelligent and What It Means for the Future*, which I authored with Dr Michael Woodley of Menie, and *The Past is a Future Country: The Coming Conservative Demographic Revolution*, which I wrote with J.O.A. Rayner-Hilles, Esq. Figures 2 and 3 originally appeared in the latter.

Edward Dutton
16th December 2022, Oulu, Finland

BREEDING THE HUMAN HERD

One

Allergic to the Basics of Life

The Case of Nicky Ratcliffe

On the night of Thursday 20th January 2000, 18-year-old undergraduate Nicola Ratcliffe was enjoying a meal with a group of friends at *Rajpooth*, an Indian restaurant just outside the centre of Durham in northeast England. The Medieval cathedral city is home to England's third oldest university and the group of first-year students—all of whom lived with Nicola's boyfriend on the same corridor in the gothic College of St. Hild and St. Bede—were celebrating their return from the long Christmas break. They were tucking into their chicken kormas, lamb dopiazas and keema naans. The drink was flowing.

The private-school girl, who was originally from near Guildford in Surrey, knew she was allergic to nuts. But she didn't know that there were any nuts in her curry. In those days, menus didn't yet highlight the presence of nuts. But it had been cooked in nut oil.[1] The talented musician, who was also studying English Literature and Psychology, knew that she was supposed to carry her "epi-pen" with her at all times, so that if she had an allergic reaction, it could be overcome by an injection of adrenaline. But, as she may already have learnt on her psychology course, the tendency to "follow the rules"—known as "Conscientiousness"—increases with age, apart from a blip in late adolescence

1 P. Crosthwaite, "Second Uni Death Hits Durham," *The Courier* (16th February 2000).

during which people become *less* conscientious,[2] leading to the culture of hedonism and promiscuity for which English universities were so notorious at the time.[3] This is why nut allergies are so dangerous among people of precisely this age: young adults fail to scrupulously "follow the rules" and their parents are no longer there to ensure that they do.[4]

It had been such a long time since Nicky had last had an allergic reaction that she had gradually stopped bothering to carry her potentially life-saving epi-pen around with her. She had left it in her room at St. Mary's College; an all-female institution on the other side of the small city centre. But, very soon after beginning her curry, Nicky knew she was having a reaction and she knew that the one thing that she desperately needed was at least a 20-minute walk away across cobbled streets and up a steep hill. As the reaction intensified, Nicky felt tightness in the throat, shortness of breath and a sensation of swelling in her mouth. This would all have created a sense of panic. To make matters worse, Nicky's blood-pressure would have significantly dropped, further inducing fear and confusion.[5]

Wheezing and frightened, Nicky began to make her way on foot back to St. Mary's College to retrieve her vital epi-pen, with her boyfriend accompanying her. Being the age that they were, however, they completely underestimated just how serious the situation was and just how little time they had. As they walked, Nicky's condition deteriorated. She collapsed in front of the Student Union building, about halfway to her destination. Nicky's boyfriend held her hand as she drifted into unconsciousness. An ambulance rushed her to hospital but it was too late. She was

2 C. Soto, O. John, S. Gosling and J. Potter, "Age differences in personality traits from 10 to 65: Big Five domains and facets in a large cross-sectional sample," *Journal of Personality and Social Psychology*, 100 (2011): 330–348.

3 For an interesting historical and sociological examination of Durham University, see J. Abbott, *Student Life in a Class Society* (Oxford: Pergamon Press, 1971).

4 M. H. Gowland, "Food allergen avoidance: Risk assessment for life," *Proceedings of the Nutrition Society*, 61 (2002): 39–43.

5 D. Foskett, P. Paskins, A. Pennington and N. Rippington, *Theory of Hospitality and Catering* (London: Hachette, 2016).

dead on arrival.[6]

A few days later, I sat next to Nicky's boyfriend and his mother—who had come up to Durham to comfort her bereft son—at "college tea." He was in a daze; stunned. I'd met Nicky at a pub called *The Swan and Three Signets* the previous December. She was pretty and out-going. So many people were in shock. The tragedy was even reported in the national newspapers.[7] How could this happen? How could it be, seemingly, increasingly common and severe, such that, at the time of writing, some British schools ban nuts from their premises?[8] They do this because a child could have a fatal allergic reaction simply by touching a surface that another child, who had eaten nuts, had recently touched.[9] How could a healthy food, widely eaten throughout human history,[10] suddenly be killing people?

The Collapse of Darwinian Selection

The answer, in part, is that people's genetic health, especially in more developed countries, has long been collapsing. Until around 1800, we were under conditions of harsh Darwinian selection for strong genetic health. Those who survived long enough to pass on their genes were ideally adapted to this environment. They had optimum immune systems, strong enough to fight off killer childhood diseases, such as measles, but not so primitive

6 M. Cottingham and A. Hebden, "Students Mourn Double Tragedy," *Palatinate* (4th February 2000).

7 *The Guardian*, "Student dies after 'allergic reaction'" (22nd January 2000); *Irish Times*, "Allergy girl dies" (22nd January 2000).

8 T. Burrows, "'The only nut ban should be the head': Parents blast primary headteacher's 'ridiculous' proposal to completely bar NUTS from school grounds," *Mail Online* (4th January 2018), https://www.dailymail.co.uk/news/article-5234339/Primary-school-Exeter-bans-nuts-school-grounds.html.

9 S. Haslett, "'He doesn't share his food': Parents turn on mum who wants to send her child to school with peanut butter and raw nuts," *Mail Online* (1st June 2021), https://www.dailymail.co.uk/femail/article-9638607/Parents-turn-mum-wants-send-child-school-peanut-butter-raw-nuts.html.

10 B. Reilly, *Disaster and Human History: Case Studies in Nature, Society and Catastrophe* (Jefferson, NC: McFarland, 2016), 303.

that they reacted to perfectly harmless foods as if they were lethal pathogens. Every generation, people who did not possess this optimally strong and subtle immune system would die as children, before they were able to procreate themselves. Those who would die as children would have "mutant genes," which almost always lead to a sub-optimally functioning body due to how finely adapted humans are to their ecology. Such people would have what is known as "high mutational load" and, every generation, illness would purge them from the population, such that the population remained genetically healthy.

In 1800, at the dawn of the Industrial Revolution, infant and child mortality was as high as 50 per cent in many European countries. Child mortality was the "Crucible of Evolution" and only the genetically "fittest"—those with the fewest mutant genes and thus with the least defective immune systems—survived childhood at all.[11] In effect, something akin to "broken genes" were being produced each generation, as distinct from variant genes that had no discernible negative impact. In order that the population did not become swamped by these broken genes, with their copying errors, the carriers of these broken genes would be selected out, particularly by childhood diseases.

With the Industrial Revolution, this all changed, and by 1870 medics were describing a phenomenon that had never been observed to any significant extent before: hay fever, an allergy to pollen.[12] Why did it change? Firstly, it had been getting warmer since the "Maunder Minimum" period of extreme cold that was reached in the late seventeenth century; a period also known as the "Little Ice Age." This meant that, as we moved away from the Maunder Minimum, a country like England could grow more food, as it had increasingly milder winters and warmer summers. This, in turn, meant that the country could sustain a larger population and, with less environmental harshness, there was less pressure to select out mutant genes, meaning that more (genet-

11 T. Volk and J. Atkinson, "Is Child Death the Crucible of Human Evolution?" *Journal of Social, Evolutionary and Cultural Psychology*, 2 (2008): 103–116.

12 T. Platts-Mills, "The Allergy Epidemics: 1870–2010," *Journal of Clinical Immunology*, 136 (2015): 3–13.

ically relatively sicklier) people survived into adulthood to pass on their genes.[13]

Secondly, the Industrial Revolution itself had numerous highly significant effects that acted to relax selection pressures for genetic health. It initially worsened health, of course, because increasing numbers of people moved into the cities, to urban slums where they were at high risk of dying from cholera or typhus.[14] In 1840, in part due to massive child mortality, the Manchester working class had a life expectancy of 17, whereas it was 40 nationwide and 44 for the better off.[15] But, ultimately, the Industrial Revolution massively increased the pace of production and, eventually, substantially raised living standards, such that the speed at which it continuously increased production was faster than the speed of population growth. This allowed the population to substantially increase, meaning that there were more and more people, with relatively high mutational load, passing on their genes—people who would not previously have lived long enough to do so.

Thirdly, and most importantly of all, the Industrial Revolution heralded crucial medical breakthroughs: vaccinations and enormous improvements in medical care and in scientific understanding. These all resulted in massive advances in public health. One of the key consequences was the collapse of child mortality, from 50 per cent in 1800 to less than 1 per cent in Western countries today.[16] In other words, the purging of mutations that had taken place intensely every generation for many centuries had, to a great extent, stopped. The eventual result would be an enormous rise in poor health that would be mainly "genetically-mediated"; that is to say, caused by genetic rather than just

13 M. A. Woodley and A. J. Figueredo, *Historical Variability in Heritable General Intelligence: Its Evolutionary Origins and Sociocultural Consequences* (Buckingham: University of Buckingham Press, 2013).

14 D. Geary, *European Labour Protest, 1848–1939* (London: Routledge, 2021).

15 J. Horn, *The Industrial Revolution: History, Documents and Key Questions* (Santa Barbara, CA: ABC-Clio, 2016), 73–74.

16 Volk and Atkinson, "Is Child Death the Crucible of Human Evolution?" *op. cit.*

by environmental factors. Alongside this, more and more people with serious genetic disorders, such that they would be unlikely to pass on their genes even now, would survive beyond infancy each generation, rendering them an ever-more conspicuous presence in society, to the extent of influencing the functioning of society which needed to expend resources and change in order to accommodate them. At the same time, however, there was a considerable reduction in disabilities that could be cured or prevented by environmental intervention, such as rickets; but, as we will see, the new disabilities were incurable, and those who suffered from them could live relatively long lives if aided in doing so.

Indeed, as we will see in this book, the situation is actually worse than this. For various reasons that we will explore, in advanced societies genetically healthy people are particularly prone to *not* passing on their genes, while those who are genetically unhealthy—up to a certain point of genetic poor health—are actually *more likely* to have large families. People who are genetically prone to heart attacks and obesity, for example, have higher fertility than those who are genetically sound.[17] In terms of mental health, people who suffer from Attention Deficit Hyper-Activity Disorder (ADHD), a significantly genetic condition, have higher fertility than do those who are mentally healthy.[18] Those who are strongly prone to taking risks and failing to plan for the future—in other words, those who have psychopathic traits, these again being highly heritable—have higher fertility than ordinary, mentally stable people.[19]

The Rise of Allergies

How does this relate to what happened to Nicola Jane Ratcliffe?

17 D. Hugh-Jones and A. Abdellaoui, *Natural Selection in Contemporary Humans is Linked to Income and Substitution Effects* (University of East Anglia, 2021).

18 N. Mullins, A. Ingason, H. Porter et al., "Reproductive fitness and genetic risk of psychiatric disorders in the general population," *Nature Communications*, 8 (2017): 1–6.

19 M. Jokela, "Birth-cohort effects in the association between personality and fertility," *Psychological Science*, 23 (2012): 835–841.

The answer is that Nicky's immune system was sub-optimal. This statement may sound harsh but if we are to make sense of something then we have to put aside emotion and coldly look at the facts.

Nicky was friendly, popular, and an excellent musician. She played the saxophone, the clarinet, and the piano, and she had been the leader of her school jazz band.[20] But it is a fact that Nicky possessed a sub-optimal immune system, almost certainly partly due to elevated mutational load. Also, under less pristine conditions, Nicky probably wouldn't have had such a strong allergy in the first place—although she was genetically allergy-prone—because her immune system would have been busy fighting off all sorts of microbes instead, meaning it would have developed in a more adaptive way, stimulated across time in the way it was evolved to be.[21] But a severe enough allergy would have been a problem even in the less hygiene-conscious conditions of the past.

Some people have an under-active immune system, one which is not strong enough to fight off infection without medical assistance. This means that they are more likely to succumb to all manner of infections, such as pneumonia or sepsis. The immune system worsens with age, meaning that older people, until quite recently, have generally died of infection, especially of pneumonia, rather than of cancer, heart attacks, or strokes. In the nineteenth century, the latter conditions were actually very rare causes of death and have risen in prevalence, in part, due to changes in diet and changes in levels of physical activity in developed societies.[22] With the rise of vaccinations for viral diseases, antibiotics for bacterial infections, and improved medical science, the chances that people with maladjusted immune systems,

20 *Surrey Live,* "Popular student made 'lasting impression'" (4th February 2000), https://www.getsurrey.co.uk/news/local-news/popular-student-made-lasting-impression-4859196.

21 See, A. Liu, "Hygiene theory and allergy and asthma prevention," *Paediatric and Perinatal Epidemiology,* 21 (2007): 2–7.

22 P. Clayton and J. Rowbotham, "How the Mid-Victorians Worked, Ate and Died," *International Journal of Environmental Research and Public Health,* 6 (2009): 1235–1253.

who are thus prone to infection, will live long and fertile lives are hugely increased. It is a slight over-simplification to speak of "weak" immune systems. More precisely, there is an imbalance within the immune system that leads to weakness with regard to certain conditions. Most readers, at the time of writing, will know someone, or know someone who knows someone, who died of COVID-19. Those who die of COVID-19—such as my office-mate when I was a postgraduate, Mennonite Pastor the Rev'd Rick Conrad (1966–2021)—don't die because of their "weak" immune systems, but because one part of the immune system, the TH1 system, overreacts and triggers a cytokine storm strong enough to kill them.[23]

Nicky Ratcliffe, in a sense, had the opposite problem. Her immune system was *too* strong or, rather, a crude and primitive part of the immune system was over-compensating for loss of function in the more sophisticated part. This is why viral infections can exacerbate allergies,[24] and it is why allergies are common in people with compromised immune systems, such as those with HIV.[25] In people with allergies, the immune system, lacking sophistication, over-reacts to what most human immune systems would realise were relatively harmless substances or a relatively harmless amount of that substance, such as the venom of a single wasp sting. The body will produce white blood cells in order to attack the substance and keep it out of the body, but if it does so to too intense a degree then the result is an allergic reaction, in which the body is effectively attacking itself. It reacts to the foreign body with such ferocity that it compromises the functioning of other parts of the body, such as by preventing the body from inhaling sufficient oxygen. An extreme allergic reaction, of the kind that Nicky experienced, is known as "anaphylactic shock,"

23 D. Ragab, H. Eldin, M. Taeimah et al., "The COVID-19 Cytokine Storm: What We Know So Far," *Frontiers in Immunology* (2020), https://doi.org/10.3389/fimmu.2020.01446.

24 P. Tantilipikorn and P. Auewarakul, "Airway allergy and viral infection," *Asian Pacific Journal of Allergy and Immunology,* 29 (2011): 113–119.

25 D. More, "Do People with HIV Get More Allergies?" *Very Well Health* (2nd February 2021), https://www.verywellhealth.com/hiv-and-allergies-82632.

and it can be deadly, as it was in her case. Asthma and eczema are also examples of auto-immune responses in which the body over-produces a particular antibody, resulting in an allergic reaction.

Unsurprisingly, in that allergies are ultimately caused by an over-active immune system, sufferers will frequently have multiple allergies. And, also unsurprisingly, with the extreme weakening of Darwinian selection pressures brought about by modern medical science, there has been an enormous rise in the prevalence of allergies in recent decades. Around 20 per cent of people in developed countries now have an allergy, and in some developed countries it is as high as 40 per cent.[26] Approximately 20 per cent of these allergy-sufferers are so allergic that there is the very real possibility that they could become seriously ill or, in the absence of necessary treatment, die of their condition at any time, due to a severe asthma attack or due to anaphylactic shock.[27] Anyone who has ever witnessed someone have a serious asthma attack will know that it is a truly horrible experience, and naturally far worse for the asthmatic, as the sufferer desperately struggles to breathe, find his inhaler, and, in effect, save his own life. The number of people admitted to European hospitals with severe allergic reactions increased seven-fold between 2005 and 2015.[28] Asthma was not reported as increasing until 1970, and peanut allergies only started to gain significant coverage in 1990.[29] A British local newspaper in 1994 headlined, "Peanut Allergy Can Kill—GPs Warned."[30]

This acceleration in cases is extremely rapid, which implies that the genetic deterioration of the immune system is likely in-

26 R. Pawankar, *The WAO White Book on Allergy* (Milwaukee, WI: World Allergy Organization, 2013).

27 EAACI, *Tackling the Allergy Crisis in Europe: Concerted Policy Action Needed* (European Academy of Allergy and Clinical Immunology, 2016).

28 B. Nwaru, L. Hickstein, S. Penasar et al., "Prevalence of common food allergies in Europe: A systematic review and meta-analysis," *Allergy*, 69 (2014): 992–1007.

29 Platts-Mills, "The Allergy Epidemics," *op. cit.*

30 *Sandwell Evening Mail*, "Peanut Allergy Can Kill—GPs Warned" (11th May 1994).

teracting with some kind of environmental change. According to the "Hygiene Hypothesis," we now live in an excessively clean and sterile environment and we are also inoculated against, and otherwise protected from, formerly common childhood diseases. This is leading to our immune system suffering from a kind of stunted development due to a lack of early stimulation; it is causing our immune system to atrophy or, at least, to not develop precisely as it should.[31] We are suffering from reduced exposure to a wide palette of pathogenic and not-quite-pathogenic microorganisms. A second possibility, though this would seem to contradict the first, is that globalization and population growth means the increased spread of, and genesis of, new, infectious diseases, of varying degrees of potency. Even if we don't realise it, our bodies are having to work extra hard to fight off this growing disease burden, leading to our immune systems developing sub-optimally.

In Britain, as of 2004, 20 per cent of the population had an allergy.[32] Allergies can be acquired throughout the lifespan, meaning that 44 per cent of British adults suffer from at least one allergy.[33] Among those under the age of 35, 50 per cent suffer from at least one allergy while, according to 2010 data, the rate is 30 per cent among those who are 65 years or older. Almost half of allergy-sufferers (48 per cent) have more than one allergy,[34] consistent with allergies being—at least in part—an expression of an underlying sub-optimal immune system. High levels of IgE (Immunoglobin E) seem to be a common denominator. IgE is the type of antibody that causes allergic symptoms, by activating mast cells (cells in connective tissue that contain histamines) and

31 E. von Mutius, "Allergies, infections and the hygiene hypothesis: The epidemiological evidence," *Immunobiology*, 212 (2007): 433–439.

32 M. L. Levy, "Inadequacies in UK primary care allergy services: National survey of current provisions and perceptions of need," *Journal of Clinical and Experimental Allergy*, 34 (2004): 518–520.

33 Mintel Press Office, "Not to Be Sneezed At: Almost Half of Brits Are Allergy Sufferers" (26th March 2010), https://www.mintel.com/press-centre/beauty-and-personal-care/not-to-be-sneezed-at-almost-half-of-all-brits-are-allergy-sufferers.

34 Mintel Press Office, "Not to Be Sneezed at: Almost Half of Brits Are Allergy Sufferers," *op. cit.*

basophils (a type of white blood cell responsible for inflammatory reactions in immune responses) in mucous membranes (the lining of organs, attacked by pathogens), for example the respiratory epithelium in asthma.

The number of people with serious allergies is also rising at an astonishing rate. Between 1992 and 2012, there was a 615 per cent increase in people being admitted to British hospitals with anaphylactic shock.[35] Between 1977 and 2007, the number of British children diagnosed with hay fever and eczema trebled.[36] Almost a third of allergy sufferers are forced to significantly change their lifestyles, including by keeping their homes extra-clean (11 per cent) or even by investing in specially-made bedding (11 per cent).[37] I know one woman, for example, who is so allergic to cats that she could not live in a house where the owner prior to the owner who was selling the house had kept a cat. Humans have co-evolved with cats—they keep our food stores free from mice in return for food and shelter,[38] and they are mood-elevating pets and companions,[39] even if they are essentially little psychopaths—so being allergic to our evolutionary partners is clearly maladaptive. Up to 10 per cent of the British population have a food allergy, while up to 20 per cent experience some kind of mild allergic reaction to some foodstuff, such as nuts or shellfish.[40] Around 7 per cent of breast-fed infants in the UK already have an allergy. Approximately 2.5 per cent of the

35 P. Turner, M. H. Gowland, V. Sharma et al., "Increase in anaphylaxis-related hospitalizations but no increase in fatalities: An analysis of United Kingdom national anaphylaxis data, 1992–2012," *Journal of Allergy and Clinical Immunology*, 135 (2015): 956–963.

36 R. Gupta, "Time trends in allergic disorders in the UK," *Thorax*, 62 (2007): 91–96.

37 Mintel Press Office, "Not to Be Sneezed at: Almost Half of Brits Are Allergy Sufferers," *op. cit.*

38 D. W. Paxton, "Dogma and Catma: Coevolution of People, Dogs and Cats," *The Humanistic Psychologist*, 49 (2021): 577–588.

39 A. Beetz, K. Unväs Moberg, H. Julius and K. Kortschal, "Psychosocial and Psychophysiological Effects of Human-Animal Interactions: The Possible Role of Oxytocin," *Frontiers in Psychology*, 3 (2012): 234.

40 The Association of UK Dieticians, "Food Fact Sheet: Food allergies and intolerances" (London, 2015).

British population are allergic to peanuts, a number which doubled between 2005 and 2015.[41] And these are increasingly joined by allergies—or "intolerances"—that were practically unknown a few generations ago, such as gluten and milk.[42]

So common have these kinds of food allergies become that there is an increasing tendency, on the part of parents organising their children's birthday parties, to assume that at least one of the children invited will have some kind of allergy, probably to nuts, and so avoid the use of nuts in the birthday cake or in any other party food. Parents will ask other parents, on the child's invitation, to "let us know" if their child has a food allergy. In 2019, the budget airline *Easyjet* announced that it was banning the sale of peanuts—once a stock in-flight snack—on all of its flights. It further announced that it would prevent passengers from eating their own peanuts, which they may have brought onto the plane, if it was made aware that there was someone with a peanut allergy on the flight. This is because an allergic passenger might inhale tiny particles from the peanuts and have an allergic reaction as a consequence. German national carrier *Lufthansa* has also stopped selling peanuts aboard its aircraft.[43] Increasingly, companies cover themselves from complaint or legal action by stating on products that do not contain nuts that they cannot guarantee that these products have not somehow been contaminated with tiny traces of nuts, such as during harvesting or transportation.[44] In the coming years, as people with nut allergies have children and as their children have children, it is probable that such allergies will become an ever-more significant problem and will become ever-more life-threatening in their intensity. And

41 BSACI, "Allergy in Children" (London: British Society for Allergy and Clinical Immunology, 2021), http://www.bsaci.org/resources/allergy-in-children.

42 R. Pomerance Berl, "Making Sense of the Gluten-Free Food Frenzy," *US News* (1st June 2012), https://health.usnews.com/health-news/articles/2012/06/01/making-sense-of-the-gluten-free-food-frenzy.

43 *BBC News*, "Easyjet bans nuts on all flights to protect passengers" (24th April 2019), https://www.bbc.com/news/business-48041464.

44 J. Barnett, K. Muncer, J. Leftwich et al., "Using 'may contain' labelling to inform food choice: A qualitative study of nut allergic consumers," *BMC Public Health*, 11 (2011): 734.

there is also a psycho-somatic component to some of these allergies, notably to peanuts—they reflect partly-genetic increasing poor mental health—which will only compound the problem.[45] In other words, some of these allergies involve a nocebo effect—the opposite of the placebo effect—where if hypochondriacs are convinced that they are allergic to peanuts then it can cause an allergic reaction.[46]

The Rise of Poor Genetic Health

However, as the human genome crumbles, due to weakened selection pressures, we would expect to see all kinds of other changes in the nature of developed populations and, indeed, we have long been seeing them. This process—of the "crumbling genome"[47]—is known as "dysgenics," a word coined by English physician Caleb Saleeby (1878–1940),[48] and meaning "bad genes." Saleeby suggested the word to the scientist Sir Francis Galton (1822–1911) and wrote about doing so in a book he published in 1914.[49] Galton coined the term "eugenics," which means "good genes," in 1883 in his book *Inquiry into Human Faculties and Their Development*.[50] Galton wrote: "The word eugenics would sufficiently express the idea; it is at least a neater word and a more generalised one than viriculture, which I once ventured to use."[51] This "dysgenic" movement away, in all directions, from what was

45 J. Kelso, C. Connaughton, R. M. Helm and W. Burks, "Psychosomatic peanut allergy," *Journal of Allergy and Clinical Immunology*, 111 (2003): 650–651.

46 See, H. Fraser, *The Peanut Allergy Epidemic: What's Causing It and How to Stop It* (London: Simon and Schuster, 2015).

47 A. Kondrashov, *Crumbling Genome: The Impact of Deleterious Mutations on Humans* (Hoboken, NJ: John Wiley & Sons, 2017).

48 R. Lynn, *Dysgenics: Genetic Deterioration in Modern Populations* (London: Ulster Institute for Social Research, 2011), 5.

49 C. Saleeby, *The Progress of Eugenics* (London: Cassell and Company, 1914), 182.

50 J. Daar, *The New Eugenics: Selective Breeding in an Era of Reproductive Technologies* (New Haven, CT: Yale University Press, 2017), 31.

51 F. Galton, *Inquiries into Human Faculty and Its Development* (London: Macmillan, 1883), 25, FN 1.

being selected for under harsh Darwinian conditions will be the focus of this book. As we will see, we are witnessing the decline of intelligence, of what used to be called "moral character" (a pro-social personality), of religiousness, of mental health, and of physical health. We are witnessing this because those were the traits that previously optimized the passing on of one's genes. With the weakening of Darwinian selection there is decreasing pressure for those traits and, in some cases, the opposite traits have ended up being positively selected for.

Having explored the evidence for each of these "dysgenic" trends, we will then look at how this potential problem was predicted in the mid-nineteenth century and at how many academics attempted to promote a policy of "eugenics." "Eugenics" is commonly defined as the study of how to arrange reproduction within a human population in order to increase the occurrence of heritable characteristics regarded as desirable. "Positive eugenics" involves actively promoting an increase in supposedly desirable traits, such as by encouraging highly intelligent people to have more children. "Negative eugenics" involves decreasing undesirable traits, such as by incentivising those with low intelligence to limit their fertility.[52]

As we will see below, the word "eugenics" has become a highly emotive and poorly-defined "academic insult" with which ideologues can attempt to besmirch research which they dislike. Rather like "racist" or, historically, "atheist" or "papist," "eugenicist" is, in the words of Australian historian Alexandra Walsham, one of these "categories of deviance to which individuals who were even marginally departed from the prescribed ideals might be assimilated and thereby reproved."[53] In other words, "eugenics" is an ill-defined term, damaging to those accused of espousing it, that is employed to shut down debate by associating academic dissidents with it, and thus with "evil" and "Nazism." As educationalist Bernadette Baker has put it, "*Eugenics* is a problematic,

52 D. Shaw, *Genetic Morality* (New York: Peter Lang, 2011), 126.

53 A. Walsham, *Church Papists: Catholicism, Conformity and Confessional Polemic in Early Modern England* (Woodbridge, Suffolk: The Boydell Press, 1999), 108.

contentious and emotive term."[54] Thus, we define "eugenics," in line with Galton's definition, as we have defined it above.[55] It is, therefore, precisely the system that a gardener would utilize when growing plants or that a farmer would employ when breeding livestock. And if we accept the Darwinian model of human origins, then humans are nothing more than a highly intelligent form of livestock, one which is, potentially, its own farmer.

Under a eugenic regime, healthy, moral and intelligent people would be encouraged to procreate and their opposite would be discouraged from doing so or, in extreme cases, even prevented from doing so. Up until World War II, "eugenics" was popular among many educated people, even, indeed especially, among those who were politically on the left such as the novelist H. G. Wells (1866–1946), and a number of eugenics policies were implemented in various countries. German genetics Nobel laureate Hermann Muller (1890–1967) was a Communist who wrote a letter to Soviet leader Joseph Stalin (1878–1953) urging him to adopt eugenic policies.[56] Muller also penned an article entitled "Our Load of Mutations" in 1950, where he presented the principle of mutation-selection balance and in which he made clear that without selection, either natural or imposed by eugenic measures, the load of bad mutations will increase from generation to generation.[57] London played host to the "International Eugenics Congress," an academic conference, in 1912.[58] In 1913, the Oxford Union Society—the Oxford University student debating society—voted by a large margin that "this house approves of the

54 B. Baker, "The Hunt for Disability: The New Eugenics and the Normalization of School Children," in J. Alan and A. Artiles, (Eds.), *World Yearbook of Education 2017: Assessment Inequalities* (London: Routledge, 2016), 138.

55 F. Galton, "Eugenics: Its Definition, Scope and Aims," *American Journal of Sociology,* 10 (1904): 1.

56 C. Sanders, *Radiobiology and Radiation Hormesis: New Evidence and its Implications for Medicine and Society* (New York: Springer, 2017), 78.

57 H. Muller, "Our Load of Mutations," *American Journal of Human Genetics,* 2 (1950): 111–176.

58 *The Scotsman,* "International Eugenics Congress," (27th July 1912).

principles of eugenics."[59]

In 1913, a eugenics education conference was held at the University of London and was attended by a thousand head teachers.[60] At this conference, J. Arthur Thomson (1861–1933), Regius Professor of Natural History at Aberdeen University, for example, spoke on "The Method of Introducing the Eugenic Ideal into Schools."[61] The following year, this campaign indirectly resulted in parents at a girls' school in Dronfield in Derbyshire refusing to send their children to school and being prosecuted as a result.[62] They were horrified that the headmistress was teaching their daughters about eugenics which was considered sexual in nature and thus inappropriate for children. Widely reported in the newspapers, this became something of a national scandal,[63] with such headlines as "Parents Revolt at Dronfield."[64] Interestingly, the suffragettes and other British early feminists were strong supporters of eugenics, arguing that only some form of feminism could save the British "race" from degeneration. They averred that intelligent women were obviously going to refuse to get married if this limited their independence and made them the property of a man, with their property automatically passing to their husband. Such women would enter professions, such as teaching or nursing, in which there was a ban on marriage. Accordingly, these intelligent women would not have children. Thus, the only way to ensure that the future of the breed was not left to unintelligent women, argued the feminists, was to overturn these sexist laws.[65] We will see that mainly due to the rise of

59 G. R. Searle, *Eugenics and Politics in Britain, 1900–1914* (Leyden: Nordhoof International Publishing, 1976), 13.

60 *Daily Citizen,* "Eugenics for the Young" (3rd March 1913).

61 J. A. Thomson, "The Method of Introducing the Eugenic Ideal into Schools," *Proceedings of the Eugenics Education Conference* (London: Eugenics Education Society, 1913).

62 *Derbyshire Courier,* "Dronfield Parents in Court" (27th June 1914).

63 P. L. Whitfield, *"The Lady Who Dared:" An Examination of the 1914 Dronfield School Strike* (Master's Thesis: Sheffield Hallam University, 2021).

64 *Daily Citizen,* "Teaching of Eugenics: Parents Revolt at Dronfield" (23rd June 1914).

65 A. T. Allen, "Feminism and Eugenics in Germany and Britain, 1900–

"individualism," the entire issue of eugenics became taboo.

But before we attend to any of this, we need to examine in greater depth the traits that were being intensely selected for until the Industrial Revolution, and why they were being selected for. And we also need to explore the results of this process: that all of the traits that were selected for have, therefore, become genetically—or "pleiotropically"—related to each other. This means that, for example, low intelligence is associated with gene forms that cause poor physical health while religiousness is negatively associated with gene forms that cause poor mental health. In the next chapter, we will make sense of why this is the case and also of the other pathways that lead to these associations.

1940: A Comparative Perspective," *German Studies Review,* 23 (2000): 477–505.

Two

What Was Being Selected for Under Pre-Industrial Conditions and How?

Jaywick, Essex

There is a town in Essex in southeast England called Jaywick. The small seaside community, of approximately 4,000 people, is about two miles from Clacton-on-Sea which, many decades ago, before most English people could afford foreign holidays, was a reasonably successful tourist resort. Jaywick itself was founded in 1928 as a cheaper version of Clacton, specifically so that working class Londoners could afford a holiday by the sea. However, with the rise, from the 1970s onwards, of cheap vacations in hotter countries, the "traditional English seaside holiday" has increasingly been consigned to history. I remember in the 1980s, as a young child, my parents and I would holiday in Charmouth and Lyme Regis in Dorset in the southwest of England, and these towns would be busy with other English holidaymakers. By the 1990s, we were going on holiday to France and Spain, as were so many others, as airlines were deregulated, permitting "no frills" flights and very cheap air travel.[1]

As this change occurred, many coastal resorts, which were financially reliant on the British summer holidaymaker, became increasingly impoverished and fell into ruin. Businesses closed, shops shut down, and there was an exodus of people away

1 See, D. Richardson, *Let's Go! A History of Package Holidays and Escorted Tours* (Stroud: Amberley Publishing, 2016).

from these areas.[2] As we will see later, "intelligence"—the ability to solve cognitive problems, a trait which robustly predicts socioeconomic status and is negatively associated with general conservatism—also predicts migration.[3] As such, overall, those who remained behind would have been those who were less intelligent. Low intelligence would make them conservative, fearful of change, low in openness to new possibilities, and also present-oriented rather than future-oriented, meaning that they would be unwilling to make sacrifices or effort "in the now" for a future reward. They would be most unlikely to "save," for example, tending, instead, to spend money on immediate gratification. Such people would also be poorly organised and would be more likely to lack the money to migrate elsewhere or even the skills necessary to attain a job elsewhere or, indeed, any job at all. Naturally, this process resulted in the collapse of property prices in these areas, causing these towns to act like magnets for slum landlords and thus for poorer people from the region who were in search of cheap places to live. As a consequence, many of these seaside resorts are now among the most deprived towns in Britain. Jaywick is the most deprived of all.[4]

The people of Jaywick are not just impoverished: they are also extremely unhealthy. This manifests in very high levels of obesity, elevated rates of multiple chronic medical conditions, alcoholism, and also in poor mental health, resulting in relatively high levels of suicide.[5] Of course, an element of this is partly explicable in environmental terms. People who are poor will not be able to afford healthy—and generally more expensive—food.

2 J. Glancy, "Britain's seaside ruins," *BBC Culture* (10th June 2015), https://www.bbc.com/culture/article/20150609-britains-seaside-ruins.

3 A. Jensen, *The* g *Factor: The Science of Mental Ability* (Westport, CT: Praeger, 1998).

4 *BBC News*, "England's most deprived areas named as Jaywick and Blackpool" (5th September 2019), https://www.bbc.com/news/uk-england-49812519.

5 V. Chalmers, "Living in a deprived neighbourhood raises the risk of an early death, being obese and drinking too much alcohol, scientists confirm in first ever review of the evidence," *Mail Online* (2nd October 2019), https://www.dailymail.co.uk/health/article-7521937/A-deprived-neighbourhood-drives-higher-risk-premature-death.html.

They will live, instead, on a diet of cheap convenience food, high in salt and sugar. However, evidence indicates that in reality, unable to budget and lacking the impulse control or even ability to cook, they live off takeaways and ready meals, both of which are more expensive than cooking from scratch. Consistent with this, some charity-run "food banks" in the UK—which give out food to those who claim they can't afford it—force people to register for a budgeting and cookery course in return for their food parcel.[6]

Unemployment may make people depressed, causing them to turn to alcohol—and even to suicide. In addition, if their poverty is mediated by low intelligence then they will have poor impulse control, causing them to drink too much, eat too much, and eat the tastiest—and least healthy—foods, all the while failing to think about the future. But, on the other hand, as we have already seen, there is a significant genetic component to this. Genes for low intelligence significantly cross-over with those associated with poor physical health, such as proneness to heart disease,[7] and poor mental health.[8]

This means that the poor are increasingly unable to cope with the adversities that can be caused by being poor, as well as unable to solve the problems that arise. They are evolved, in effect, to an environment of plenty, where there is no struggle for survival, and so no selection against genetic mutations that lead to poor health and low intelligence. As we will explore later, many of them probably wouldn't have survived childhood under even slightly harsher conditions than we live in today. No wonder the so-called "Cost of Living Crisis," which started to be discussed in 2022, led to emotional dysregulation and suicidal ideation among such

6 B. Glaze, M. Smith, R. Wearmouth and J. Hudspith, "Tory MP slammed for claims people using food banks 'can't budget' and 'don't know how to cook,'" *Daily Post* (12th May 2022).

7 I. Deary, S. Harris and W. D. Hill, "What genome-wide association studies reveal about the association between intelligence and physical health, illness, and mortality," *Current Opinion in Psychology*, 27 (2019): 6–12

8 S. Bahrami, A. Shadrin, O. Frei et al., "Genetic loci shared between major depression and intelligence with mixed directions of effect," *Nature: Human Behaviour*, 5 (2021): 795–801.

people.[9] Suicide attempts may possibly be understood in evolutionary terms as a "cry for help" which manipulates others into assisting, as most attempts do not end in death. Suicide may also be a by-product of an adaptive propensity, especially in unstable yet easy ecologies, to feel negative feelings strongly, one of which is depression,[10] but in the wrong ecology this leads to suicidal ideation. We will discuss such ecologies in more depth below. But why are intelligence and poor health genetically associated?

What is Pleiotropy?

The answer is via a process called "pleiotropy." Put simply, if two traits are both being selected for in an organism—for example, intelligence and physical health—then, gradually, they will stop varying independently of each other, which means that they will start to become "correlated;" that is, "associated."

"Pleiotropy" describes the observation that most genetic variants have more than one kind of effect. They can affect multiple organ systems, because the expression of most genes is widespread in the body. For example, when a mutation reduces the immune system's ability to fight certain infections (because the mutation happens to be in a gene that is important for immune defences), there is a good chance that this gene is also expressed in tissues other than the those relating to the immune system and that in these other tissues it is used for purposes other than fighting infections. Therefore, the functions of these other tissues

9 K. Rowlett, "Cost of living crisis pushing people to consider suicide warns Somerset charity," *ITV News* (22nd June 2021), https://www.itv.com/news/westcountry/2022-06-21/cost-of-living-crisis-is-leaving-people-suicidal-warns-somerset-charity. On the relationship between emotional dysregulation and suicidal ideation, see: H. Turton, K. Berry, A. Danquah and D. Pratt, "The relationship between emotion dysregulation and suicide ideation and behaviour: A systematic review," *Journal of Affective Disorders: Reports*, 5 (2021): 100136.

10 There are other possibilities. Altruistic people may become suicidal if they consider themselves a burden on their kin, for example. See, H.-J. Aubin, I. Berlin and C. Kornreich, "The Evolutionary Puzzle of Suicide," *International Journal of Environmental Research on Public Health*, 10 (2013): 6873–6886.

will also be impaired by the mutation. Thus, pleiotropy can affect any combination of phenotypic traits, but similar functions, especially functions requiring the same tissue type, are more often affected together. For example, the expectation is that mutations that reduce intelligence are more likely to also cause psychopathology than they are to cause, for example, muscle weakness or infertility. But, put briefly, this is why genetically adaptive traits tend to end up bundled together, as do genetically maladaptive traits. This is why we can talk of there being a "fitness factor." In addition, around 84 per cent of the genome relates to brain functioning, rendering the brain a massive target for mutation. It follows that if a person carries mutations of the body, causing a poor immune system for example, he will very likely carry mutations of the mind, causing poor mental health.[11]

Natural Selection

But before we look in greater detail at how this works, we must familiarise ourselves with how selection operates. *Natural Selection* is the process whereby an organism is adapted—both physically and mentally—to a particular kind of ecology. A gene is composed of two alleles. In the process of procreation, with most genes, an infant inherits one allele from each parent. Sometimes something will go wrong, and one of these genes will not copy correctly. This leads to a defective gene, a "mutation." Under the brutal conditions of Darwinian selection, organisms are fine-tuned to their environment. Therefore, any mutation that has an effect on gene function at all will almost always make the gene less functional, and the organism less well-adapted to its environment. Each child is born with anywhere from 40 to 80 new mutations, depending on the age of the father. Most are in the realm of junk DNA and cause no harm, but an unknown fraction causes mild impairment, and a few are bad enough to cause a disease or disability, such as Huntington's Disease or Achondro-

11 See, M. Sarraf, M. A. Woodley of Menie and C. Feltham, *Modernity and Cultural Decline: A Biobehavioral Perspective* (London: Palgrave Macmillan, 2019).

plasia, which leads to dwarfism.[12]

Accordingly, the carriers of these mutations will be selected against, because they are more likely to die young, or are less likely than others to have many children even if they survive. Sometimes, a mutation will actually give the carrier an advantage in its particular environment. Pale skin, for example, would have given the first pale-skinned mutant an edge in a cloudy northern climate because he or she would have been able to absorb more sunlight and convert it into Vitamin D.[13] When this kind of mutation manifests in the right kind of environment, the allele will be selected for because the carrier will be less likely to be disabled by rickets and more likely to end up with a large number of descendants. However, in general, mutations are selected against.

We all inherit mutations from our parents under our current conditions of weakened selection pressure. Each of us carries about three new mutations which neither of our parents carried. In most cases, these don't do us much harm.[14] It has also been estimated that the genome in modern Western countries, specifically in Iceland, acquires at least 60 new mutations every generation.[15] Iceland is a useful place to conduct such studies due to the Icelandic being a small, isolated and homogenous population with universally accessible healthcare.[16] However, the rise in genome sequencing in recent years has meant that these results have been found in many countries, not just Iceland.[17] This process, of new mutations manifesting every generation, has two key consequences.

12 R. M. Pauli, "Achondroplasia: A Comprehensive Clinical Review," *Orphanet Journal of Rare Diseases*, 14 (2019): 1.

13 See G. Cochran and H. Harpending, *The 10,000 Year Explosion: How Civilization Accelerated Human Evolution* (New York: Basic Books, 2009).

14 M. W. Nachman and S. L. Crowell, "Estimate of the mutation rate per nucleotide in humans," *Genetics*, 196 (2000): 297–304.

15 A. Kong, M. Frigge, G. Masson et al., "Rate of de novo mutations and the importance of father's age to disease risk," *Nature*, 488 (2012): 471–475.

16 J. Gulcher, A. Helgason and K. Stefansson, "Genetic homogeneity of Icelanders," *Nature Genetics*, 26 (2000): 395.

17 See, Sarraf, Woodley of Menie and Feltham, *Modernity and Cultural Decline, op cit.*

Firstly, there are simply more carriers of genes that cause severe defects and, thus, more people who suffer from these severe defects, such as Huntington's Disease, which we will examine below. But secondly, it means that almost all human genomes now contain both small and large defects. In some cases, genetic diseases are not due to having a specific faulty gene, as in Huntingdon's Disease, but as a consequence of *locus heterogeneity*, meaning that a deleterious mutation in any one of a number of genes can lead to the increased risk of a disease. This is the case with Alzheimer's for example.[18] The many flaws that almost every human being has—both physical and mental—are to a large extent caused by the assortment of genetic defects that they carry. These defects work together to make every aspect of the carrier work sub-optimally, resulting in increases in disease susceptibilities, intellectual limitation, minor physical defects, and minor mental defects. Under Darwinian conditions, most of these mutations would have been purged every generation due to high child mortality, although mutations that caused only a very minor limitation could survive and become common in a particular area with a small, homogenous population due to "founder effect" (one fertile founder heavily influences a small population) or "genetic drift" (when allele frequencies simply change in chance ways across time). In general, though, such mutations were selected out. But, as we have discussed, this process has broken down—due to medical advances, inoculation, better living conditions and cheaper food—leading to an increasing "mutational load" with each new generation.

How Do Mutations Arise?

Mutations arise as copy errors during DNA replication, either with or without damage to the DNA. They can be expressed in four ways, because there are four kinds of genetic variants, or alleles: additive, recessive, dominant and X-linked. *Dominant alleles* affect the trait which they impact when the allele is present in one copy besides the alternative "normal" allele, because they

18 L. Hernandez and D. Blazer, (Eds.), *Genes, Behaviour and Social Environment* (Washington, DC: National Academies Press, 2006), Ch. 3.

dominate the normal allele. *Recessive* alleles are expressed only when they are present in two copies. *Additive*, or intermediate, inheritance means that the effects of the alleles add up. Most complex traits, such as height, blood pressure and intelligence, are *polygenic*. This means that natural variations in them are caused by many different genes working together to produce the relevant trait. As a highly polygenic trait, intelligence is easily disrupted by mutations because mutations in a very large number of different genes, a few thousand of them, can interfere with its expression.

Epistastic Effects: Bipolar Disorder

The effects of some mutant alleles are only triggered if another mutant allele in a different gene is also inherited. This is known as *epistasis*. Many studies have found that certain psychiatric disorders, such as schizophrenia and bipolar disorder, sometimes manifest due to the *epistatic effects* on a complex trait of two different alleles both being present in one individual.[19] This would mean that both parents could be psychologically normal, though, between them, carry both of the alleles that lead to an epistatic effect if they are both inherited.

For example, a mutation in one gene may put its carrier at risk of manic-depressive (bipolar) disorder, but *only* if that person also carries a mutation in a *different* gene. If both mutations happen to be inherited by a child, the child will grow up to develop bipolar disorder. This is a form of depression characterized by oscillating between periods of extreme happiness and energy ("mania") and periods of profound despair. A famous sufferer is the English comedian and writer Stephen Fry (b. 1957). He attempted suicide in 2012 and had previously come close to it a number of times.[20]

19 C. Webber, "Epistasis in Neuropsychiatric Disorders," *Trends in Genetics*, 33 (2017): 256–265.

20 S. Cable, "A downcast Stephen Fry appears the day after he reveals he tried to commit suicide last year with a 'huge a number of pills and vodka,'" *Mail Online* (5th June 2013), https://www.dailymail.co.uk/news/article-2336406/Stephen-Fry-tried-commit-suicide-year-large-number-pills-vodka-saved-producer-unconscious-body.html.

Recessive: Sickle Cell Anemia

There can also be a significant effect if you inherit two mutant *recessive alleles* of the same gene, one from each parent, so that no normal copy of the gene is present. Sometimes, this is sufficient to cause specific genetic disorders. For example, research in northern Finland, where there are elevated rates of mild mental retardation compared to the rest of the country, has highlighted a defect in a gene known as *CRADD*, a defect which is 50 times more frequent in Finland than elsewhere in Europe. It is a recessive allele, and if both parents carry a copy of the defective *CRADD* allele then there is a 25 per cent chance that any child will be mildly mentally retarded through inheriting two copies of it.[21] Carrying one copy of the allele does no harm because, being recessive, it is not expressed unless the other allele is present.

There are, however, cases where a genetic disease stays in the population because inheriting one copy of the mutant allele benefits health but inheriting two copies damages health. For example, people who suffer from sickle cell anemia, which is most prevalent among Sub-Saharan Africans, have two copies of the same mutant allele in a gene encoding hemoglobin. However, those who carry only one copy are not only healthy, but have improved immunity to malaria.[22] Improved malaria resistance has been strongly selected for in human populations that were exposed to this disease in the past, while the offspring with sickle cell anemia would rarely have lived long enough to reproduce. This is because sickle cell anemia is a life-threatening condition, the symptoms of which include a low number of red blood cells, recurrent episodes of vascular occlusion with severe pain and tissue infarction, frequent infection, delayed growth, delayed puberty, and vision problems in some patients.[23] American Jazz

21 M. Kurki, E. Saarentaus, O. Pietiläinen et al., "Contribution of rare and common variants to intellectual disability in a sub-isolate of Northern Finland," *Nature Communications,* 10 (2019): 410.

22 L. Luzatto, "Sickle cell anaemia and malaria," *Mediterranean Journal of Hemotology and Infectious Diseases,* 4 (2012): e2012065.

23 See, P. Jones, *Genes and Sickle Cell Disease* (New York, Chelsea House, 2008).

musician Miles Davis (1926–1991), who suffered from sickle cell anemia, died of pneumonia, which had plagued him throughout his life, aged 65.[24]

Dominant: Huntingdon's Disease

By contrast, some conditions arise from a single mutant *dominant allele*. Dominant alleles dominate in the sense that a single mutant allele, when paired with the normal one, will always cause a disease. Huntington's Disease is a neurodegenerative disorder that starts in young or middle-aged adults with subtle behavioral changes, then progresses to a severe movement disorder with involuntary overshooting movements accompanied by progressive dementia, problems swallowing, speaking, and breathing, and death about 20 years after the onset of the first symptoms.

Symptoms usually start to manifest between the ages of 30 and 50. It is a dominant condition. If one of your parents has Huntington's disease, this parent will have a normal and a mutant allele, and you will have a 50-50 chance of inheriting either the normal or the mutant allele.[25] The American folk musician Woody Guthrie (1912–1967)—writer of "This Land is Your Land," among other hits—died of Huntington's Disease aged 55. He inherited the condition from his mother.[26] However, one might, of course, develop such a condition due to a *de novo* mutation passed on by a parent, which happens in 10 per cent of cases of Huntington's Disease, usually via the father.[27]

X-Linked: Hemophilia and Turner Syndrome

X-linked alleles are the exception. They do not come in pairs.

24 J. Szwed, *So What? The Life of Miles Davis* (New York: Random House, 2012).

25 See, T. Visser, *Huntington's Disease: Etiology and Symptoms, Diagnosis and Treatment* (Hauppage, NY: Nova Science Publishers, 2010).

26 K. Coombs, *Woody Guthrie: America's Folk Singer* (Minneapolis, MN: Carolrhoda Books, 2002).

27 G. Houge, O. Bruland, I. Bjørnevoll et al., "De novo Huntington disease caused by 26–44 CAG repeat expansion on a low-risk haplotype," *Neurology,* 81 (2013): 1099–1100.

They are on the X chromosome, and they are passed from parent to child. A chromosome is an aggregate of genes all of which, together, control a series of functions. For example, the sex of an organism is dictated by it having either two X chromosomes (female) or an X and a Y chromosome (male). A normal human has 23 chromosomes. One of the most well-known X-linked conditions is the blood-clotting disorder hemophilia, which affected many of the inter-married royal families of Europe in the nineteenth and twentieth centuries, specifically the British, German, Spanish, and Russian Royal Houses. Hemophilia renders the blood very slow to clot, resulting in excessive bleeding. It appears that hemophilia must have entered these royal families via the British Royal Family as a *de novo* mutation on the sperm of Prince Edward, Duke of Kent (1767–1820). Prince Edward was one of the younger sons of King George III (1738–1820). As a result of this mutation, Prince Edward's daughter, Queen Victoria (1819–1901), was a "carrier" of hemophilia.[28]

With X-linked conditions, the mother is a "carrier" but never displays any symptoms. Her daughters have a 50 per cent chance of carrying the disease, of which they will display no symptoms, and a 50 per cent chance of being normal. Her sons have a 50 per cent chance of *suffering* from the disease and a 50 per cent chance of being normal. If these male hemophiliacs have children, then they will pass on the hemophilia mutation only to their daughters, who will all be carriers, as the affected male will always pass on one X chromosome to a daughter. This is what occurred with Queen Victoria's hemophiliac son Prince Leopold, Duke of Albany (1853–1884). Prince Leopold's daughter, Princess Alice, Countess of Athlone (1883–1981), was a carrier.[29] More importantly, Queen Victoria's daughter, Princess Alice, Grand Duchess of Hesse (1843–1878), was a carrier, as was Alice's own daughter, Alexandra, Tsarina of Russia (1872–1918). The result was that the Russian Crown Prince, Alexey (1904–1918), suffered from hemophilia, an issue which destabilized Russia's absolute mon-

28 D. M. Potts, *Queen Victoria's Gene: Haemophilia and the Royal Family* (Stroud: History Press, 2011).

29 R. K. Massie, *Nicholas and Alexandra: The Classic Account of the Fall of the Romanov Dynasty* (New York: Random House, 2000), 147.

archy, contributing to revolution, civil war, and the takeover of Russia by the Bolsheviks.[30]

Sometimes, one of the two X chromosomes is not passed on to the female or it is only partially passed on. When this happens, the female develops Turner Syndrome. This is marked by sterility as well as by a number of physical abnormalities including a short and wide neck, a broad chest with widely spaced nipples, short stature, limited or no sexual changes at puberty, low-set ears and downward-slanting eyes. Sufferers have normal intelligence, but they tend to have impairments in aspects of visio-spatial skills and mathematics.[31]

Paternal Age and Mutation

Mutations are more likely to stem from the father and there may manifest as many as 200 new mutations per generation, though perhaps as few as 40.[32] Having an older father raises the risk of new mutations because the spermatogonia, the stem cells from which the sperm cells are formed, undergo many rounds of cell division in a man's life, with a risk of new mutations every time the genome replicates and is distributed to the daughter cells.[33] This reflects a broader process whereby we all acquire so called "somatic mutations" as we get older, due to our cells copying with decreasing accuracy. In addition, as we age, reduced effectiveness of DNA repair makes us more vulnerable to the low-level background radiation to which we are continuously exposed.[34] Some chemical elements are unstable, which means

30 D. M. Potts, *Queen Victoria's Gene: Haemophilia and the Royal Family* (Stroud: History Press, 2011).

31 P. Fechner, *Turner Syndrome: Pathophysiology, Diagnosis and Treatment* (New York: Springer, 2020).

32 Y. Xue, Q. Wang, Q. Long et al., "Human Y Chromosome Base-Substitution Mutation Rate Measured by Direct Sequencing in a Deep-Rooting Pedigree," *Current Biology,* 19 (2009): 1453–1457.

33 S. Wu, F. Wu, Y. Ding et al., "Advanced parental age and autism risk in children: a systematic review and meta-analysis," *Acta Psychiatrica Scandinavica,* 135 (2016): 29–41.

34 J. Tong and T. Hei, "Aging and age-related health effects of ionizing radiation," *Radiation Medicine and Protection,* 1 (2020): 15–23.

that their nuclei decay, releasing energy in the form of radiation. The immune system doesn't fight ionizing radiation. But we have DNA repair systems that repair the damage that the radiation does to the DNA. Also, the older you are the more oxidizing "free radicals"—atoms with unpaired electrons that increase your exposure to radiation—you have in your body, which damage the DNA and thereby cause mutations.[35]

All of this causes our bodies to age, and so function less well.[36] Somatic mutations will not affect your offspring, of course, but mutations in the spermatogonia will, and these, like the somatic mutations, will accumulate with age. It is likely no coincidence, in this regard, that Queen Victoria's father was past the age of 50 when she was conceived, significantly elevating the probability of *de novo* mutations, including haemophilia. We also inherit *de novo* mutations from our mothers, though to a much lesser extent, and the likelihood of inheriting them also increases with maternal age. One study found that with two parents in their late teens or early twenties, a child would inherit 9 new mutations from their mother, but 39 *de novo* mutations from their father. By the time the parents were aged 40, the child would, on average, inherit 15 new mutations from the mother but 91 from the father.[37]

Maternal Age and Down's Syndrome

Even so, people sometimes inherit seriously health-damaging conditions via their mothers. One of the most well-known disorders associated with maternal age is Down's Syndrome. Older eggs are more likely to fail to properly divide. The consequence of this can be the child inheriting a partial or full extra copy of Chromosome 21, with Down's Syndrome being the result. If the

35 V. Lobo, A. Patil, A. Phatak, and N. Chandra, "Free radicals, antioxidants and functional foods: Impact on human health," *Pharmacognosy Review*, 4 (2010): 118–126.

36 R. A. Risques and S. R. Kennedy, "Aging and the rise of somatic cancer-associated mutations in normal tissues," *PLOS Genetics*, 14 (2018): e1007108.

37 Kong, Frigge, Masson et al., "Rate of *de novo* mutations, father's age and disease risk," *op. cit.*

child inherits a full extra copy, then the condition will be more severe than if it inherits a partial extra copy. In about 2 per cent of cases, the child has two copies in some cells and three in others. This particularly mild form of the condition is known as "Mosaic Down's Syndrome."[38]

Officially known as "Mongolism" until the World Health Organization changed it to "Down's Syndrome" in 1965, the disorder is named after English physician John Langdon Down (1828–1896), who was the first person to describe it in detail.[39] It is characterised by cognitive impairment of varying degrees, a distinctive physical appearance (including a flattened face, small head, short neck, protruding tongue, slanting eyes [hence "Mongolism"], misshapen ears, poor muscle tone, a simian crease on the palm, and reduced height), heart defects, gastrointestinal defects, immune disorders, sleep apnoea, obesity, and early onset dementia, among many other health problems.[40] As a consequence, in 1947, the life expectancy of a British child born with Down's Syndrome was 12 years—though it has since risen to between 50 and 60 years due to medical advancements.[41] Genetic mental problems are often correlated with an usual appearance, with both being underpinned by broader disruption to developmental pathways: the subtler or milder the mental problem; the subtler the physical abnormalities.[42]

Down's Syndrome children, at least, are also known for their friendly, caring and pro-social personalities, in comparison to

38 G. Thomas, *Down's Syndrome Screening and Reproductive Politics: Care, Choice, and Disability in the Prenatal Clinic* (London: Routledge, 2017).

39 O. Ward, "John Langdon Down: The Man and the Message," *Down Syndrome Research and Practice,* 6 (1999): 19–24.

40 See, R. Newton, *The Down's Syndrome Handbook: The Practical Handbook for Parents and Carers* (New York, Random House, 2011).

41 J. Carr, *Down's Syndrome: Children Growing Up* (Cambridge, Cambridge University Press, 1995), 4.

42 A. Lee, C. Hibbs, M. Wright, M. et al., "Assessing the accuracy of perceptions of intelligence based on heritable facial features," *Intelligence,* 64 (2017): 1–8. For an examination of physiognomy see: E. Dutton, *How to Judge People By What They Look Like* (Oulu: Thomas Edward Press, 2018).

those of their age.[43] This may be a result of them aging quickly,[44] with people tending to become higher in empathy as they get older.[45] The result of this accelerated aging is that many people with Down's Syndrome already suffer from dementia by the time they reach middle age. About 2 per cent of people with Down's Syndrome have dementia by the age of 30, with 50 per cent of them suffering from dementia by the time they reach the age of 60.[46] This compares to 2 per cent of the general population who suffer from dementia at the age of 60.[47]

The English actress Sally Phillips (b. 1970)—famous for playing the lead role's frequently-swearing friend in the *Bridget Jones* films—had her first child, a son, when she was 35. This rendered her, medically, *almost* a "geriatric mother." A "geriatric mother" is defined as a mother over the age of 35. Beyond the age of 35, there is a significantly elevated risk of maternal-age-related disorders in the child, though this risk increases with age even throughout the 30s. The actress' son was diagnosed with Down's Syndrome when he was 10 days old. Phillips subsequently had two healthy boys, evidencing the way in which maternal age merely increases the *probability* of the offspring suffering from genetic health problems.[48] When a woman is 25, there is a 0.08

43 D. Fidler, "The emergence of a syndrome-specific personality profile in young children with Down syndrome," *Down Syndrome Research and Practice*, 10 (2006): 53–60.

44 W. Zigman, "Atypical aging in Down Syndrome," *Developmental Disabilities Research Reviews,* 18 (2013): 51–67.

45 C. Soto, O. John, S. Gosling, and J. Potter, "Age differences in personality traits from 10 to 65: Big Five domains and facets in a large cross-sectional sample," *Journal of Personality and Social Psychology,* 100 (2011): 330–348.

46 A. Wray, *The Dynamics of Dementia Communication* (Oxford, Oxford University Press, 2020), 269, FN 5.

47 R. George-Carey, D. Adeloye, K. Y. Chan et al., "An estimate of the prevalence of dementia in Africa: A systematic analysis," *Journal of Global Health*, 2 (2012): 02041.

48 E. Buckland, "'It was a shock but our lives are more meaningful': Sally Phillips discusses her first child Olly's diagnosis with Down's syndrome and praises her son's 'jokes and humour,'" *Mail Online* (19th August 2020), https://www.dailymail.co.uk/tvshowbiz/article-8638545/Sally-Phil-

per cent chance of her child being born with Down's Syndrome. By the time she is 40, the risk has risen to 1 per cent and by the time she is 45 it is 3 per cent.[49] These likelihoods, of course, relate to the *average* woman, in health terms. It took Phillips two years to become pregnant with her first son and she struggled through multiple miscarriages to have her subsequent two sons.[50] Maternal fertility problems predict having a Down's Syndrome child,[51] possibly due to maternal elevated mutational load, with an unhealthy woman's eggs less likely to function correctly.

In addition, there is a hereditary form of Down's Syndrome called "Translocation Down's Syndrome" where "carriers," who display no symptoms, have an extra Chromosome 21 fused with one of their other chromosomes. When this fusion chromosome is transmitted together with a normal chromosome 21, the child has Down's Syndrome because it has all of the genes of chromosome 21 in three rather than two copies. Female carriers are more likely to pass on the condition and maternal age has no bearing on whether they do so.[52] Males with Down's Syndrome are often infertile due to spermatogenesis defects, though there have been cases of males with non-Mosaic Down's Syndrome fathering children.[53] Females with Down's Syndrome can more easily have children, but they experience premature menopause,[54] on average at the age of about 45, around six years younger than healthy,

lips-discusses-child-Ollys-diagnosis-Downs-syndrome.html.

49 A. Olsen, D. Redfearn and A. Pepe, *A-Z of Learning Disability* (London: Macmillan, 2016).

50 Buckland, "'It was a shock but our lives are meaningful,'" *Mail Online, op. cit.*

51 H. Shalaby, "A study of new potential risk factors for Down syndrome in Upper Egypt," *Egyptian Journal of Medical Human Genetics,* 12 (2011): 15–19.

52 G. Thomas, *Down's Syndrome Screening and Reproductive Politics: Care, Choice, and Disability in the Prenatal Clinic* (London: Routledge, 2017).

53 M. Pradhan, A. Dalal, F. Khan, and S. Agrawal, "Fertility in men with Down syndrome: a case report," *Fertility and Sterility,* 86 (2006): 1765.

54 E. Parizot, R. Dard, N. Janel, and F. Vialard, "Down syndrome and infertility: what support should we provide?" *Journal of Assisted Reproduction Genetics,* 36 (2019): 1063–1067.

European females.[55]

Sexual Selection

The second form of selection highlighted by Charles Darwin (1809–1882) was *Sexual Selection.* In most animal species, males will fight to mate with as many females as they can. In winning these fights, they establish who among them is the strongest, who is the healthiest and who, therefore, has very few mutant genes. Females are sexually attracted to males who are successful in these fights. The dominant male, as far as the females are concerned, will provide them with genetically healthier children and these children are more likely to survive the rigours of natural and sexual selection themselves. This will mean that the female's own genes are more likely to be passed on. The population will remain healthy and strong, in part because those who lack genetic health—those with a high percentage of mutant genes—will find that the females do not want to have sex with them. The females will aggressively fight off any attempt by an unattractive male—a male who is high in mutational load—to breed with them. In modern societies, rape will generally involve a desperate and unattractive male having to force a relatively attractive woman to have sex with him, because there's no other way of obtaining sex, apart from prostitution.[56]

Darwin himself observed that, "It is certain that among all animals there is a struggle between the males for possession of the female."[57] Throughout animal species, males compete for territory or, in the case of the social animals such as chimpanzees, the group competes for territory, but each male competes for a position within the status hierarchy of the group. Only those who are successful in accruing territory or attaining high status within the hierarchy of the group will be attractive to the females.

55 M. Cosgrave, J. Tyrell, M. McCarron, M. Gill and B. Lawlor, "Age at onset of dementia and age of menopause in women with Down's syndrome," *Journal of Intellectual Disability Research,* 43 (1999): 461–465.

56 See R. Thornhill and C. Palmer, *A Natural History of Rape: Biological Causes of Sexual Coercion* (Cambridge, MA: MIT Press, 2000).

57 C. Darwin, *The Descent of Man* (London: John Murray, 1871), II, 239.

And the males will ultimately attract the females through winning fights, attaining status, or displaying other qualities that indirectly betoken healthy genes.

In addition, the females will generally be specifically attracted to the kinds of qualities that lead to the males obtaining status in the first place. In a society where status is essentially achieved through fighting, these will be markers of physical strength and genetic health: muscles; size; a nice furry mane as an indicator of the healthy genes necessary to successfully grow one. Males will showcase these desirable qualities through fighting in front of the females, but also by what is known as "strutting." They will walk around advertising their genetic quality. An example of this is the peacock's tail. As American psychologist Geoffrey Miller has noted, only a genetically healthy peacock will have sufficient resources left over to be able to grow a large and colourful tail. A less healthy peacock will have to invest more of his bio-energetic resources in fighting off disease, leading to the production of a gaudy, asymmetrical or small tail.[58] Chimpanzees will deliberately display their erections, conveying to the females that they are genetically healthy enough to be able to direct resources into growing a large penis.[59] With human males, the secondary sexual characteristic of musculature implies a good immune system, such that sufficient resources are left over to be directed towards muscle protein synthesis. Testosterone is actually negatively correlated with immune system activity, further rendering testosterone manifestations, such as musculature, expressions of genetic health.[60] In addition, the human male will stereotypically display his (supposed) wealth and, by implication, his relatively high place in the hierarchy. He has the resources left over for the designer shirt and the sports car, or, at least, he wants the females

58 This point has been made in, G. Miller, *The Mating Mind: How Sexual Choice Shaped the Evolution of Human Nature* (New York: Anchor Books, 2000).

59 A. Dixson, *Primate Sexuality: Comparative Studies of the Prosimians, Monkeys, Apes, and Humans* (Oxford: Oxford University Press, 2012).

60 W. Lassek and S. Gaulin, "Costs and benefits of fat-free muscle mass in men: relationship to mating success, dietary requirements, and native immunity," *Evolution and Human Behavior,* 30 (2009): 322–328.

to think that he does.

As we will discuss in more detail below, in all human societies intelligence, especially combined with a socially effective kind of personality, predicts socioeconomic status, and both traits are strongly genetic.[61] Indeed, we have evidence that intelligence was being selected for across time until at least the early nineteenth century due to its association with socioeconomic status, something that meant that wealthy men were extremely attractive to women and that their offspring were simply more likely to survive.[62] This was possibly caused, in part, by richer women being better fed and having a lower physical workload than poorer women. This would also speed up the resumption of fertile cycles after birth and shorten inter-birth intervals.

Passing on Your Genes Indirectly

The third form of selection highlighted by Darwin is *Social Selection.* This relates to how a group treats its various members and the impact this has on whether they pass on their genes. For example, does the group execute most of its criminals or not? If it does, it will be acting to remove whatever genes relate to criminality from the gene pool, so this will be social selection in action. When two groups that adopt different forms of social selection come into conflict, then there will be *Group Selection*, with one group tending to triumph over the other.

It should be noted that there is considerable debate over the utility of "group selection" as a construct. American evolutionary biologist David Sloan Wilson[63] has advocated the *Multi-Level Selection Theory*. He argues that once cooperative groups develop within a species, selection will act to promote those groups that possess the optimum level of those qualities that permit them to

61 See, R. Lynn, *Dysgenics: Genetic Deterioration in Modern Populations* (London: Ulster Institute for Social Research, 2011).

62 See, E. Dutton and M. A. Woodley of Menie, *At Our Wits' End: Why We're Becoming Less Intelligent and What It Means for the Future* (Exeter: Imprint Academic, 2018).

63 D. Wilson and E. Sober, "Reintroducing group selection to the human behavioural sciences," *Behavioural and Brain Sciences*, 17 (1994): 585–654.

out-compete other groups. Thus, selection will still operate on individuals within a group but can also be seen to operate on groups themselves, as collections of individuals (who identify as such and create group borders) and, in some circumstances, can shift away from individual and towards group selection. This model, as we will now see, helps to explain, for example, the development of altruistic tendencies. It can be conceived of as a logical extension of something that most evolutionary biologists happily accept, *Kin Selection.*

British biologist William Hamilton (1936–2000) conceived of what he called "inclusive fitness."[64] This is the idea that there are a number of ways in which you can pass on your genes; it doesn't have to be only through direct descendants. Hamilton argued that we all follow a process of "kin selection" where we invest in people who are closely related to us. Thus, uncles and aunts will invest in their nieces and nephews, with whom they share 25 per cent of their genes, in context where they compare themselves to other members of their ethnic group. That is, if two random co-ethnics are understood to not be related, then an aunt and a nephew are 25 per cent genetically the same. Cousins will help each other out, sharing 12.5 per cent of their genes, and this may even extend to cousins once removed, with the older one treating the younger one as a *de facto* niece or nephew. This process is why the spinster aunt, who has no children of her own, will invest in and give presents to her siblings' children. In doing so, she is not simply being altruistic, because if she were then she could just give money to charity. She is helping to pass on her own genes, a quarter of which are carried by each nephew or niece. For this reason, in some circumstances, it would make sense, in terms of inclusive fitness, to lay down your life if a large number of your cousins were under mortal threat, especially if you had already had children yourself. According to Hamilton, people will act altruistically if the benefit to their inclusive fitness is greater than the fitness cost of the act; something known as "Hamilton's Rule." Thus, it would make sense for a menopausal

64 W. D. Hamilton, *The Narrow Roads of Gene Land* (Oxford: Oxford University Press, 1996); W. D. Hamilton, "The genetical evolution of social behaviour: I and II," *Journal of Theoretical Biology*, 7 (1964): 1–52.

mother to lay down her life to save her only child. This would not make so much sense if the mother, aged 21, was told to make a choice between her unborn baby's life and her own, as she could go on to have many more children. In these circumstances, we would expect her to choose her own life.

"Group selection" can be regarded as an extension of kin selection. It is a genetic fact that ethnic groups are genetic clusters. The average Englishman is highly genetically similar to the next average Englishman *relative* to the average Dane, based on genetic assay data. Australian political psychologist Frank Salter has calculated that if the world were divided between only English and Danes, then the relationship between two average English people would have a kinship coefficient of 0.0021, where it would be zero for an Englishman and a Dane. This coefficient would be the equivalent of sharing a set of 6 time great grandparents; that is, being 7th cousins. When the genetic distance is greater, the incentive to fight is greater. If 10,000 Danes migrated to England, this would be the equivalent of every English person losing 167 children, due to the extent, for example, to which the English and Danes would integrate, and fewer English genes would be passed on. If this were 10,000 Bantus, who are far more genetically different from the English than are the Danes, then each English person would lose the genetic equivalent of approximately 10,854 children. In a world of Englishmen and Bantu, two random Englishmen are not 7th cousins; they are brothers or father and son. So, from a genetic perspective, it could be adaptive for an Englishman to fight to protect his ethnic group from Danes, even if it risked his having no children at all. If his actions saved enough of his own people, this would more than compensate for the lack of direct breeding.[65] The soldier who laid down his life in this way would be operating at the level of group selection. Computer models have shown that the more ethnocentric group, the group whose members are more inclined to repel outsiders and make sacrifices for the good of the group, always eventually

65 F. Salter, *On Genetic Interests: Family, Ethnicity and Humanity in an Age of Mass Migration* (New Brunswick, NJ: Transaction Publishers, 2007); F. Salter, "Estimating Ethnic Genetic Interests: Is It Adaptive to Resist Replacement Migration?" *Population and Environment,* 24 (2002): 111–140.

dominates in the battle of group selection, all else being equal. This group triumphs over groups composed of members who will cooperate with the enemy but not the in-group members, cooperate with everyone or cooperate with nobody.[66]

So group selection is simply a logical extension of kin selection, as ethnic groups are extended kinship groups, as has been demonstrated with genetic assay data.[67] The important qualification to this model is that people will only engage in altruistic behaviour if the genetic payoff outweighs the sacrifice. This means that it is more difficult for them to be exploited by selfish free-riders. They are altruistic, but only within certain limits. In a sense, altruism could only evolve because the mechanisms of group selection (in-group favouritism and ethnocentrism) restricted altruism to the in-group, which in the evolutionary past meant people with whom one is very familiar, which reduced the risk of being exploited. "Group selection," nevertheless, raises criticism from certain researchers. One of the reasons for this seems to be little more than a misunderstanding of what is meant by it. When modern evolutionary psychologists talk about "group selection" they mean "Multi-Level Selection" or "New Group Selection." Advocates of the original model of group selection argued that altruism developed because more co-operative groups were more likely to survive. The major weakness with this model was "subversion from within." Any group of altruists would be exploited by selfish free-riders, who would manifest by genetic chance, and these would then outbreed the altruists.[68] However, this is not what New Group Selection is arguing. As we have already noted, New Group Selection is simply extending kin selection to a very distant level of kin: the ethnic group. It is distinguishing between "kin" and "ethnic group" because most people would not necessarily conceive of their ethnic group as genetic kin. So there is a nuanced difference between New Group Selection and

66 R. Hammond and R. Axelrod, "The evolution of ethnocentric behaviour," *Journal of Conflict Resolution*, 50 (2006): 1–11.

67 Salter, *On Genetic Interests, op. cit.*

68 S. Okasha, "Biological Altruism," in E. Zalta, (Ed), *The Stanford Encyclopedia of Philosophy* (2013), https://plato.stanford.edu/archives/fall2013/entries/altruism-biological/.

kin selection. The mechanisms of reciprocity evolved perhaps not so much for interactions with strangers, but as a mechanism to permit in-group altruism and cooperation while preventing exploitation.

Refuting Criticisms of Group Selection

"Group selection," meaning "New Group Selection," has been criticized in depth by Canadian psychologist Steven Pinker.[69] His key criticisms are that: (1) "Group Selection" deviates from the "random mutation" model inherent in evolution. (2) We are clearly not going to be selected to damage our individual interests, as group selection implies. (3) Human altruism is self-interested and does not involve the kind of self-sacrifice engaged in by sterile worker bees.

Each of these points can be answered. Firstly, if the group selection model is building on the individual selection model, then it is bound to present a slightly different metaphor. To dismiss it on these grounds seems to betoken nothing more than a fervent attachment to the original metaphor. Secondly, the group selection model merely suggests that a group will be more successful if there is genetic diversity, meaning that an optimum percentage of its members are inclined to sacrifice themselves for their group. Thirdly, it is clearly the case that a small percentage, in many groups, is indeed prepared to sacrifice itself for the group, sometimes damaging their individual genetic interests. So, it seems to me that it is reasonable to accept multi-level selection. Furthermore, as we have already seen, computer-modelling has shown that groups that are internally cooperative but harsh to members of other groups—those that are high in positive and negative ethnocentrism—tend to eventually win in the battle of group selection. They always dominate the computer grid after many generations when they battle with those who cooperate with nobody, cooperate with everybody, or only cooperate with members of another group.[70] Of course, this is not necessarily

69 S. Pinker, "The false allure of group selection," *The Edge* (18th June 2012), www.edge.org/conversation/the-false-allure-of-group-selection.

70 Hammond and Axelrod, "The evolution of ethnocentric behaviour,"

the only means by which one group can dominate another. A group could be lower in ethnocentrism than another group but much higher in average intelligence, resulting in superior warfare tactics and a genius caste with outlier high IQ that innovated brilliant weapons. However, all else being equal, it is the most ethnocentric group which triumphs.[71] And in general, social selection is simply the effect that the culture of the group has on somebody's chances of survival.

It is also worth noting the language that Pinker employs in criticising group selection. He refers to it as having a "false allure," terms it a "dust bunny" and a "hairy blob" that apparently "bleeds outwards to a motley collection of other, long-discredited versions." It has, claims Pinker, "placed blinkers on psychological understanding by seducing many people." This is extremely emotionally manipulative language that attempts to compel the reader to associate "group selection" with being sexually manipulated, with stupidity, with beasts of burden, and with the word "discredited." It is the "poisoning of the well" fallacy, and the fact that Pinker engages in this would tend to imply cognitive dissonance; that he rejects group selection for his own personal reasons, realizes that the model is likely reasonable, and is thus triggered to react in an emotional fashion.

"Group selection" has also been criticized by American physicist Gregory Cochran, who claims it is vanishingly unlikely to ever manifest; "almost impossible."[72] However, it can be responded that given the millions of different species on Earth it would have come about eventually, and if it is the simplest explanation for available data on how humans behave, then it very probably has done. Group selection is real and necessary to explain evolution, but its mechanisms are not yet fully understood.[73] Even

op cit.

71 See, E. Dutton and B. G. Charlton, *The Genius Famine* (Buckingham: University of Buckingham Press, 2015).

72 G. Cochran, "Group Selection (and homosexuality)," *West Hunter* (13th January 2013), https://westhunt.wordpress.com/2013/01/10/group-selection-and-homosexuality/.

73 For further exploration see: B. G. Charlton, "Reconceptualizing the metaphysical basis of biology: a new definition based on deistic te-

so, how could positive and negative ethnocentrism ever evolve without group selection? It wouldn't make sense.

Finally, it has been argued that early human groups were too small and sparse for group selection, but this has been comprehensively refuted by abundant evidence of genocide among such groups, with genocide levels increasing with adoption of agriculture, and separated clans seeming to form large groups in order to repel invaders from different tribes.[74] Moreover, detailed modelling has found that group selection is a reality, with researchers concluding that: "If culturally transmitted systems of rules (institutions) that limit individual deviance organize cooperation in human societies, then it is not clear that any extant alternative to cultural group selection can be a complete explanation."[75] Clearly, there are no logical criticisms of group selection.

How Does It All Come Together?

So we know that, as we will explore in greater depth below, certain highly genetic traits—such as intelligence, a pro-social personality, physical health and mental health—were all being selected for under conditions of harsh Darwinian selection. The result of this was pleiotropy: that the traits have become genetically correlated. In other words, a new genetic variant that has favourable effects on two phenotypic traits is more likely to get fixed in the population than one that has a good effect on one trait and a bad effect on the other trait. Conversely a bad mutation is more likely to be bad for multiple phenotypic traits because genes are expressed in multiple places in the body. Therefore "good traits" end up going together, as English psychologist Charles Spearman (1863–1945) argued.[76]

leology and a hierarchy of organizing entities," *The Winnower*, 6 (2016): e145830.07350

74 See B. Kiernan, *Blood and Soil: A World History of Genocide and Extermination from Sparta to Darfur* (New Haven, CT: Yale University Press, 2007).

75 P. Richerson, R. Baldini, A. Bell and K. Demps, "Cultural group selection plays an essential role in explaining human cooperation: A sketch of the evidence," *Brain and Behavioural Sciences*, 39 (2016): e30.

76 See, I. Deary, F. Spinath and T. Bates, "Genetics of Intelligence," *Euro-*

To give some examples, females sexually select for intelligence in males because intelligence is associated with the ability to attain social status. They also sexually select for height because, in the unstable ecology of prehistory, a taller man could better defend them. Also, height—like the peacock's tail—implied low mutational load: a male with high mutational load would have to use disproportionately more of his bio-energetic resources to fight off disease, resulting in failure to attain his maximum phenotypic height. As a consequence, height and intelligence are weakly genetically correlated in males.[77]

Intelligence can be conceived of as rather like a pyramid. At the base of the pyramid there are numerous "specialized abilities" that are only very weakly associated with intelligence, such as the ability to tie your shoelaces or catch a ball. Moving up, there are often considered the three main forms of intelligence, which are more strongly associated with how smart you are: mathematical, verbal, and spatial intelligence. People vary on how highly they score on tests of these. Some people are more verbally-tilted, while others are more mathematically-tilted. But, in general, those who score high on one kind of intelligence score high on them all, so we can conceive of a factor which underpins all of them. This is "general intelligence," known as *g*, which is at the pinnacle of the intelligence pyramid. "General intelligence," the most genetic aspect of intelligence, is weakly correlated with "General Factor of Personality," a construct which refers to a socially effective personality of the kind that would have been selected for in harsh conditions.[78] We will explore the nature of "personality" in greater detail later. And we have already observed that intelligence is associated with genes that predict physical and mental health.

Thus, in effect, it can be argued that there is a kind of "General

pean Journal of Human Genetics, 14 (2006): 690–700.

77 M. Keller, C. Garver-Apgar, M. Wright et al., "The Genetic Correlation between Height and IQ: Shared Genes or Assortative Mating?" *PLOS Genetics,* 9: (2013): e1003451.

78 J. Aitken Schermer and P. Vernon, "The correlation between general intelligence (*g*), a general factor of personality (GFP), and social desirability," *Personality and Individual Differences,* 48 (2010): 187–189.

Fitness Factor." Researchers have theorized that a "General Fitness Factor" exists, due to the inter-correlated nature of the assorted components of genetic fitness. Its existence explains a number of paradoxes.[79] For example, certain scientists have predicted that, with its low level of selection pressure, human populations should currently be experiencing "mutational meltdown," where a population is so overwhelmed by mutational load that it simply dies out. That this is not happening would be consistent with the genome being—to a significant extent, though obviously not entirely—free from mutation in about 1800, and this could only be the case if the different kinds of mutation were inter-related via a "fitness factor." This relationship occurs because the product of a particular gene may affect more than one trait or because one gene's product belongs to a control loop, such as one regulating the timing of developmental processes, which indirectly impacts other traits. The result is that traits that are selected for together will become genetically related. This is what has happened, with varying degrees of strength, to many of the traits for which we were subject to harsh selection under pre-industrial conditions.

English geneticist Sir Ronald Fisher (1890–1962) argued that under sexual selection, mate choice (what he calls "initial noticing") will cause unrelated pairs of traits, such as height and IQ, to become genetically correlated when the sexual selection pressure acts over a long period of time. The result of this is a latent "fitness factor," of physical and mental abilities and health which serves as an index of genotypic quality.[80] Invariably, some people—and some populations—will be higher in this "fitness factor" than others.

79 Sarraf, Woodley of Menie and Feltham, *Modernity and Cultural Decline, op. cit.*, 202.

80 See, R. Schedinger, *The Mystery of Evolutionary Mechanisms: Darwinian Biology's Grand Narrative of Triumph and the Subversion of Religion* (Eugene, OR: Wipf and Stock, 2019).

The People of 1800

Once this "fitness factor" develops, then sets of correlated traits will "signal" genetic fitness to other members of the species and it will be adaptive to be attracted to these, rendering these traits all the more robustly connected. People with these traits will also have more surviving offspring, further increasing the genetic connection between these different adaptive and significantly genetic traits. Accordingly, people living prior to the Industrial Revolution had long been under intense selection for intelligence, pro-social personality, mental health, and physical health, with all these traits being predicted to be correlated both genetically and phenotypically. Genetic correlation has been demonstrated for some of them through genomic research, as we will see below.

With the gradual weakening of Darwinian selection pressures since about 1800, we would expect to see a huge spike in population—due to people having tended to have many children to compensate for child mortality and there being a lag between child mortality dropping and people absorbing this and so limiting their fertility—with this new population being increasingly composed of the relatively genetically unfit. As we will see, other factors have augmented this process. But this new population, roughly 90 per cent of which would not have survived into adulthood under pre-Industrial conditions (based on what national populations were at that point), would be expected to deviate in all directions from the population of 1800 due to their greater genetic diversity, this diversity being specifically caused by damaging low-frequency genetic variants that arose recently as new mutations and that didn't get removed by Darwinian selection pressures. This would be especially obvious in terms of intelligence, average personality and related inclinations, and physical and mental health. As noted above, these would be inter-related due to pleiotropy and the brain being a huge mutational target. We will now set out the evidence for this process having occurred.

Three

The Rise of Intelligence

The Mechanized Spinning Frame

Richard Arkwright (1732–1792) was the son of an impoverished tailor from Preston in Lancashire. In the year in which Arkwright was born, England was an agricultural economy, having changed little for centuries. By the time he died, at the close of the eighteenth century, England was industrializing. Arkwright was a significant part of the reason for this change. In 1769, he patented a mechanized "spinning frame." This was a water-powered machine which, in combination with semi-skilled labourers and the use of cotton, managed to enormously speed up the production of yarn.

Due to the innovation made by this one-time barber and self-made entrepreneur, out went the slow and inefficient weavers, hunched over their cottage spinning wheels, taking a long time to produce tiny quantities of yarn. In came huge textile factories and the ability to produce textiles at an astonishingly fast pace. This indirectly set off other innovations, and before long England was in the swing of an Industrial Revolution.[1] Why did this dramatic change take place when it did? The answer is that we had long been under intense Darwinian selection for intelligence. Eventually, the highly intelligent members of the population—the "smart fraction"—were so intelligent that they developed the

1 R. S. Fitton, *The Arkwrights: Spinners of Fortune* (Manchester: Manchester University Press, 1989).

innovations of the Industrial Revolution. This led, very quickly, to dramatic changes in human populations that weakened selection pressures for intelligence. But to understand how this happened, we first need to understand what "intelligence" is.

What is Intelligence?

"Intelligence" is defined as "the ability to reason, plan, solve problems, think abstractly, comprehend complex ideas, learn quickly, and learn from experience."[2] It is the ability to solve complex problems and to solve them quickly. The quicker you can solve a problem, then the more intelligent you are. The harder the problem has to be before it is simply beyond you, then the more intelligent you are. Intelligence is quantified by the intelligence quotient (IQ), a score that can be derived from a number of standardized tests. These tests measure reasoning ability across a wide range of areas, for example, verbal, mathematical, and spatial. Ability in each of these subsections positively correlates with ability in the other subsections, allowing us to posit that they together measure an underlying factor called "general intelligence" (*g*). This is the essence of intelligence, and it is generally what we are talking about when we say that one person is more "intelligent" than another.

It cannot possibly be argued that intelligence is somehow "a very Western concept." Proxies for intelligence include general knowledge, something which is valued in all cultures.[3] Intelligence is negatively associated with criminality, which is disliked in all cultures. Intelligence is also robustly correlated with education, income, and health. As such, intelligence cannot be dismissed as context-dependent or as only relevant in the West, nor can it be dismissed as unimportant. A full list of the traits associated with intelligence can be seen in Table 1.[4]

2 L. Gottfredson, "Editorial: Mainstream Science on Intelligence," *Intelligence*, 24 (1997): 13–24.

3 D. Buss, *The Evolution of Desire: Strategies of Human Mating* (New York: Basic Books, 1989).

4 This table was first published in E. Dutton and M. A. Woodley of Menie, *At Our Wits' End: Why We're Becoming Less Intelligent and What It*

Table 1. Variables Correlated with General Intelligence[5]

Positive Correlation		Negative Correlation
Achievement motivation	Marital partner, choice of	Accident proneness
Altruism	Logical abilities	Acquiescence
Analytic style	Media preferences	Aging quickly
Abstract thinking	Memory	Alcoholism
Artistic preference and ability	Migration (voluntary)	Authoritarianism
Atheism	Military rank	Conservatism (of social views)
Brain size and size of head	Moral reasoning and development	Crime
Craftwork	Motor skills	Delinquency
Creativity	Musical preferences and abilities	Dogmatism
Diet (healthy)	Myopia	Falsification ("Lie" scores)
Democratic participation (voting, petitions)	Occupational status	Hysteria (versus other neuroses)
Educational attainment	Occupational success	Illegitimacy
Eminence and genius	Perceptual abilities	Impulsivity
Emotional sensitivity	Piaget-type abilities	Infant mortality
Extra-curricular attainments	Practical knowledge	Racial prejudice
Field-independence	Psychotherapy, response to	Reaction times
Height	Reading ability	Religiousness
Health, fitness, longevity	Social skill	Smoking
Humour, sense of	Socioeconomic status of origin	Single/young motherhood
Income	Socioeconomic status achieved	Truancy
Interests, depth and breadth of	Sports participation at university	Trust (lack of)
Involvement in school activities	Supermarket shopping ability	Weight/height ratio
Leadership	Talking speed	
Linguistic abilities (including spelling)	Trusting nature	

Means for the Future (Exeter: Imprint Academic, 2018), 11–12.

5 These are sourced mainly from A. R. Jensen, *The g Factor: The Science of Mental Ability* (Westport, CT: Praeger, 1998), 75.

What are IQ Tests?

Intelligence is measured by IQ tests. We know that these are reliable because their results correlate with other intuitive measures of cognitive ability, such as educational success. IQ is defined as having a population mean of 100 and a standard deviation (SD) of 15, which means that your intelligence is quantified relative to others from the same population group, which is typically those who live in your country. Intelligence, overall, increases with age up until middle age, after which it gradually declines. Even a below average 16-year-old performs better than a very clever 3-year-old who can already read. IQ is therefore calculated relative to people of the same age. The average person scores 100; anything less than this is below average; and anything above it is above average. Rather like height, IQ is "normally distributed" on what looks like a bell curve. Most people have an IQ of around 100 with the percentages with lower or higher IQs tapering off on both sides. As one standard deviation (SD) is already defined as 15 points, 68 per cent of people have an IQ between 85 and 115, and 95 per cent of people have an IQ between 70 and 130. This is the "normal" range. If you score below it, you are classified as "retarded." If you score above it, then you are exceptionally bright.

Figure 1: The Bell Curve

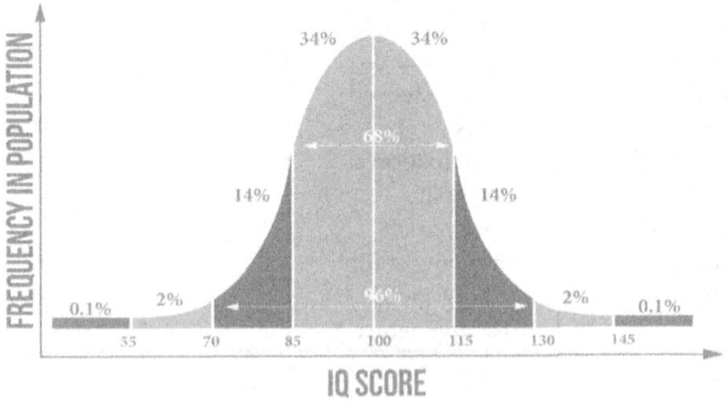

What about specific cognitive abilities? We all know people who have very good language comprehension but poor mathematical ability. But these people are unusual. Across a large group of people, all specific cognitive abilities are positively correlated with each other, and most people are either above average or below average pretty much across the board. This fact is what lays the foundation for the *g* factor (general intelligence). The validity and reliability of IQ tests has been subject to criticism, and even the concept of "intelligence" has been questioned.

A very popular alternative was proposed by American psychologist Howard Gardner.[6] Gardner argued that there are "multiple intelligences," such as bodily-kinesthetic (i.e., dancing ability), emotional, musical, and interpersonal. However, the essence of Gardner's model is that these abilities are independent, as they are related to different parts of the brain, and that a person can be high in one or more "intelligences" while being below average in others. This is simply not the case. Higher IQ people perform better in most cognitive domains, and lower IQ people perform worse. Gardner's assorted kinds of "intelligence" are either a misuse of the word "intelligence" or they are simply examples of narrow cognitive abilities underpinned by "intelligence" as normally defined. "Emotional intelligence" is a concept that has also been launched as a competitor to IQ or to "general intelligence." This ability to deal with other people and solve social problems is, however, weakly positively correlated with IQ at about 0.3.[7] The more intelligent you are then the more empathetic you are; the better you can imagine what it is like to be someone else.

Criticisms of the Validity of IQ Tests

Many criticisms have also been leveled against IQ tests but none of them really work. IQ tests have been found to have high predictive validity for school achievement (and, thus, other measures of cognitive ability), occupational status, and criminality (negative-

6 H. Gardner, *Frames of Mind: The Theory of Multiple Intelligences* (New York: Basic Books, 1983).

7 S. Kaufman, C. DeYoung, D. Reiss and J. Gray, "General intelligence predicts reasoning ability for evolutionarily familiar content," *Intelligence*, 39 (2011): 311–322.

ly).[8] They cannot be argued to be substantially culturally biased, as they correlate with objective measures, such as reaction times (negatively) and cranial capacity (positively): they correlate with how quick you are to respond to stimuli and simply how big your brain is; unsurprisingly, the brain being a thinking muscle. The robust negative correlation with reaction times—the quicker you react to a stimulus, such as a light being switched on, the cleverer you are—implies that intelligence can substantially be explained by a high-functioning nervous system.[9]

One very common argument against the validity of IQ tests is the fashionable concept of "stereotype threat." The idea is that people who belong to a group believed to do badly on IQ tests do badly because of their own expectations that they will do badly, which presumably stresses them out, reducing their performance. However, meta-analyses have found that this effect is mostly non-existent, and when it can be found—for certain people in certain situations—it is much smaller than the group differences it is used to explain away. Some studies have found that groups told that they will do badly on IQ tests, because of some factor about them, seem to consequently do *better than expected*. There is also a huge problem of publication bias in this area, with studies disproving stereotype threat simply not being published.[10] In addition, the theory does not at all address the critical question of how systematically incorrect stereotypes could originate, as 75 per cent of racial stereotypes have been found to be at least partly accurate and 50 per cent completely accurate.[11] The simplest explanation, as has been empirically explored in detail by American psychologist Lee Jussim, is that stereotypes develop because they are broadly true.[12]

8 A. Jensen, *The g Factor: The Science of Mental Ability* (Westport, CT: Praeger, 1998).

9 Jensen, *The g Factor, op. cit.*

10 C. Ganley, L. Mingle, A. Ryan et al., "An examination of stereotype threat effects on girls' mathematics performance," *Developmental Psychology*, 49 (2013): 1886–1897.

11 W. Helmreich, *The Things They Say Behind Your Back: Stereotypes and the Myths Behind Them* (New York: Doubleday, 1982).

12 L. Jussim, *Social Perception and Social Reality: Why Accuracy Dom-*

The Heritability of IQ

Meta-analyses of twin studies have found that IQ is highly genetic; it is in the region of 0.8 heritable in adult samples.[13] This means that 80 per cent of individual differences in intelligence are a product of genetics, at least in the modern Western societies where the studies have been done. The heritability of IQ is much lower among child samples because the child's environment is created not by the child but by the parents, and therefore reflects the parents' IQs, not that of the child. When the child grows up, he will start creating his own environment, one that is reflective of his own innate IQ. Accordingly, his IQ will rise or fall depending on how much more or less his "genotypic intelligence" is than that of his parents. The key environmental component to intelligence is intellectual stimulation. If you are maximally intellectually stimulated—spending time with intelligent people, having intelligent conversations, and attending a good school—then your IQ will be pushed to its "phenotypic limit."

A child who is adopted by upper middle-class parents might do very well at school because he is pushed to his maximum phenotypic intelligence by his highly able adoptive parents. However, once he goes to university—and is more in control of his own life—he will start to create an environment more in line with his own innate intelligence; likely closer to that of his low socioeconomic status biological parents.[14] Accordingly, his IQ will start to fall, perhaps to the extent that he drops out of university. But—apart from unlikely genetic combinations or epistasis causing a child to be far more or far less intelligent than his or her parents—a person's IQ is, generally, relatively similar to that of their parents. And, in general, people sexually select for those who are relatively similar in IQ to themselves, partly in order to

inates Bias and Self-Fulfilling Prophecy (Oxford: Oxford University Press, 2012).

13 R. Lynn, *Dysgenics: Genetic Deterioration in Modern Populations* (London: Ulster Institute for Social Research, 2011), 101.

14 See, E. Dutton and G. Madison, "Why do middle-class couples of European descent adopt children from Africa and Asia? Some Support for the Differential-*K* Model," *Personality and Individual Differences,* 130 (2018): 156–160.

maximize the extent to which their genes are passed on.[15] When people tend to socialize with those of their own social class and socioeconomic status, this is significantly a function of intelligence.

Selection for Intelligence Under Pre-Modern Conditions: Hunter-Gatherers

Now we are clear on the nature of intelligence, we can turn to the evidence of it having been selected for in pre-industrial societies. Sexual and natural selection for intelligence and health is very obvious in the least developed human social organizations: hunter-gatherers. These people mainly live off foraging from the local environment and meat brought into the community by male hunters.[16] These groups are unstable, often splitting into separate groups over minor arguments, and they have very high levels of child mortality. Around 50 per cent of children die from accidents or poor resistance to disease. In addition, in many of these tribes, such as the Yanomamö of Venezuela, unhealthy or simply unwanted infants will be killed by their mothers or allowed to starve to death.[17] For all of these reasons, in the groups from which modern humans evolved, there is selection in favour of the appropriate level of strength, aggression, and good genetic health.

When looking for somebody with whom to have a long-term relationship—where children might result, for example—females are more interested in a male's social status (and by extension his intelligence) than vice versa. This is likely because a person of high status will, in general, be more able, and more willing, to invest resources in the female and her offspring. She will benefit from this investment when she is pregnant and has young

15 See, J. P. Rushton, "Ethnic Nationalism: Evolutionary Psychology and Genetic Similarity Theory," *Nations and Nationalism*, 11 (2005): 489–507.

16 S. B. Eaton and M. J. Konner, "Review of Paleolithic nutrition revisited: A twelve-year retrospective on its nature and implications," *European Journal of Clinical Nutrition*, 51 (1997): 207–216.

17 See, N. Chagnon, *Yanomamö: The Fierce People* (New York: Holt, Rinehart & Winston, 1968).

children, meaning that she and the children will be more likely to survive. Accordingly, the female (and her offspring) will be more likely to pass on their genes. The children will also inherit the traits that have caused the father to achieve status such that any sons are more likely to achieve high status themselves and thus pass on their genes in a system of polygamous mating where females select for status. For this reason, there are sex differences in the qualities that we find attractive in partners. Although there are all kinds of shades of grey in between, overall, it really is true that men go for looks and women go for money and status or the potential to achieve these.[18] Men go for looks because they have less to lose from the sexual encounter. Thus, unless they intend to invest in the family, their best strategy is to have sex with as many healthy and fertile—and thus good-looking and young—women as possible.[19]

In order to better understand this selection for intelligence, let's focus on the Yanomamö or, as other tribes term them, "the Fierce People." They are a group of about 35,000 people living in around 250 separate villages on the border between Venezuela and Brazil. The "headman" will have about three wives and roughly eight children, while lower ranking men will have one or even no wives and a smaller number of children.[20] This means that only the strongest, healthiest, and most intelligent will make a significant impact on the gene pool. However, in an unstable and easy ecology, aggression and impulsivity would be useful, meaning there would be a limit on the extent to which intelligence was selected for. Nevertheless, within those limits, it is the strongest, healthiest, and most intelligent men whose genes live on. As Darwin put it, in describing these kinds of tribes:

> The strongest and most vigorous men, those who could best defend and hunt for their families, and, in later times, the chiefs or headmen, would have succeeded in leaving a greater number of offspring than would the weaker, poorer and lower members of the same

18 Buss, *The Evolution of Desire, op. cit.*

19 Buss, *The Evolution of Desire, op cit.*

20 Chagnon, *Yanomamö, op. cit.*

> tribes. The chiefs of nearly every tribe throughout the world succeed in obtaining more than one wife.[21]

This can also be seen among the Bushmen of the Kalahari, who live in small bands and migrate from one watering hole to another. One study of this group found that infant mortality was 45 per cent, and that 62 per cent of adult males produced no children at all. The headman will have an excess of wives,[22] despite the large excess of females due to male deaths in hunting accidents.[23] Reviews of research on hunter gatherers all agree that most of these peoples follow status-based polygamous mating systems.[24]

American anthropologist Janet Siskind explained in her ethnography of the Sharanahua of Peru that: "Prestige accrues to the generous hunter. Prestige is not a vague goal at Marcos, it brings a definite reward, the possibility of gaining women as lovers and/or wives. The successful hunter is usually the winner in the competition for women."[25] Such abilities would also impress the males whom they would hope to lead. But, clearly, skill at hunting would involve not just good physical health but also intelligence. The ability to hit the target or make the best weapons would be predicted by spatial intelligence and fast reaction times. Building and maintaining the weapons would require high impulse control and planning for the future. Reaching the top of the hierarchy would require the social skill involved in making alliances and understanding the best time to pose a challenge, while good verbal intelligence would be necessary to persuade and "talk down" rival males in tests of wit. So, we would expect the headmen to have relatively high intelligence by the standards of the tribe, and it would be these very people who would be disproportionately passing on their genes. American

21 C. Darwin, *The Descent of Man* (London: John Murray, 1871), 368.

22 N. Howell, *Demography of the Dobe !Kung* (New York: Academic Press, 1979).

23 P. Frost, *European Hair, Eye, and Skin Color* (Washington, DC: Academica Press, 2022), 23.

24 See G. P. Murdock, *Ethnographic Atlas* (Pittsburgh, PA: University of Pittsburgh Press, 1967).

25 J. Siskind, *To Hunt in the Morning* (Oxford: Oxford University Press, 1973), 95–96.

evolutionary psychologist David Buss has summarized that, "In tribal societies, the headman or leaders are inevitably among the most intelligent in the group."[26]

Pastoralists

Pastoralists are nomadic peoples who keep a variety of domesticated animals such as chickens. They frequently migrate from place to place to find fresh pastures for their herds. They are a more complex form of society than that of hunter gatherers; they reflect a greater degree of specialization, and there are clearer differences in social status within these groups. These differences strongly impact whether or not people have children, and we have already noted that social status is partly predicted by intelligence.

The first pastoralists appeared in Neolithic times, around 10,000 years ago, in the so-called "Fertile Crescent" on the shores of the Mediterranean. Pastoralists can sustain larger populations than hunter-gatherers because they can produce a surplus of food with which to feed these bigger populations. A modern example of such a people is the Rendille camel herders of Northern Kenya. At the age of around 11, a boy undergoes a brutal rite of passage, in public, in order to transition from being a boy to being a warrior. He must remain perfectly still as he is circumcised. If he betrays the slightest emotion, then he will shame his family and himself. At best he will be an outcast who cannot marry, and, at worst, he will be killed by his furious and humiliated relatives. Clearly, this would select for physical prowess and thus good health but also for impulse control and thus, possibly, intelligence. Once the male graduates to warrior status, he can get married, but only once he can pay the "bride price," which is paid in camels. Only about 50 per cent of Rendille males ever accrue enough camels to be able to pay the bride price, and so, in general, only the more socioeconomically successful males will father children at all.[27] We have already seen that socioeconomic

26 D. Buss, *The Evolution of Desire, op. cit.*

27 E. Moran, *Human Adaptability: An Introduction to Ecological Anthropology* (Belmont, CA: Duxbury Press, 1979).

status is predicted by intelligence, so this system would have the indirect effect of preventing those of both low intelligence and poor health from having children.

A number of these pastoralist peoples have settled down and become "agriculturalists." This means that there is even greater surplus and so an even larger population, though there will tend to be a greater concentration on one particular crop. There are hundreds of studies of these kinds of people and almost all concur that there are polygamous mating systems through which the wealthier males achieve the highest fertility. In addition, there are arduous rites of passage which weed out the males who are low in physical health. Accordingly, there is selection for intelligence and physical vigour. We would, in fact, expect selection for intelligence to be stronger among agriculturalists than among hunter-gatherers or pastoralists. To pursue agriculture successfully, you require far higher impulse control and a much stronger degree of orientation towards the future, such that you can achieve a surplus, ready for times of famine. As you are settled, and cannot move with the seasons, you have to keep warm and thus manufacture appropriate clothes and more complex and versatile dwellings. In order to marry, you must be highly successful in agriculture, being able to pay the "bride price," and this would require far higher intelligence—as reflected by the heightened ability to plan and practice self-discipline—than would hunting ability.[28]

Early Nation States

As agriculture improved further, such a large food surplus could be achieved that there was no need for everybody to even work in agriculture. Increasing numbers of people were able to pursue specialisms that had nothing to do with agriculture directly or simply employ others to work on their land while they lived a life of ease. As such, around 5000 years ago, we begin to see the development of city states, especially around the Fertile Crescent. A social class system develops. At the top, there are the kings, the

28 See: G. Cochran and H. Harpending, *The 10,000 Year Explosion: How Civilization Accelerated Human Evolution* (New York: Basic Books, 2009).

nobility, and the leading religious and military figures; then there are the wealthy farmers, merchants, and craftsmen; then there are labourers and poorer farmers; and, at the very bottom, are slaves who are literally owned by members of the higher classes.

These societies were generally polygamous, and there is sound evidence that members of the higher social classes enjoyed many wives and extremely high fertility. King Rehoboam of Judah, for example, had 18 wives and 60 concubines. Among the Incas of Peru there were legal regulations on the extent of polygamy. Emperors could have as many wives and concubines as they desired, while military officers were permitted between 15 and 30, depending on their rank.[29] In China, emperors maintained harems of hundreds of women, who would be rotated according to their time in the menstrual cycle. As such, they would have hundreds of children.[30] It is recorded that the Moroccan emperor Moulay Ishmael the Bloodthirsty (1634–1727) fathered 888 children. These were produced by hundreds of concubines and 9 wives.[31]

Jewish Eugenics

The issue of historical eugenic practices among the Jews has been explored in depth by American literature scholar John Glad (1941–2015) in his book *Jewish Eugenics*[32] and by English psychologist Richard Lynn in his book *The Chosen People*.[33] In essence, the community encouraged and funded rabbis, its most intelligent members, to have large families. Diaspora Jews were forced into high-IQ professions by law, such as banking. They also argue that "marrying-out," especially for males, was strongly frowned upon, forcing less than highly intelligent males into

29 L. L. Betzig, *Despotism and Differential Reproduction: A Darwinian View of History* (Hawthorne, NY: Aldine, 1986).

30 M. Dickemann, "The ecology of mating systems in hypergynous dowry societies," *Social Science Information*, 18 (1979): 163195.

31 M. Daly and M. Wilson, *Sex, Evolution and Behavior* (Boston: Willard Grant Press, 1983).

32 J. Glad, *Jewish Eugenics* (Washington, DC: Wooden Shore, 2011).

33 R. Lynn, *The Chosen People: A Study of Jewish Intelligence and Achievement* (Augusta, GA: Washington Summit, 2011).

infertility each generation or forcing them to effectively become gentile. Highly intelligent gentile males, evidently, sometimes married into the Jewish community, with their children becoming Jewish.

It should be noted, however, that the idea that a Jew inherently required a Jewish mother did not become Jewish Law until the second century.[34] Even so, there is evidence of this system of matrilineal descent not having always been followed, though transgression, for example in Medieval Spain, was rare and socially unacceptable.[35] But, nevertheless, consistent with the Rabbinical dictates having, indeed, been followed to a significant extent, Jewish matrilineal DNA diversity is much lower than Jewish patrilineal DNA diversity and, moreover, Jewish matrilineal DNA diversity specifically is much lower than that found among gentiles from the same regions. In fact, there is an extreme "founder effect" for Jews in the female, but not male, line. This, of course, implies that "marrying out" was much more unacceptable for females, whereas gentile males might sometimes "marry in."[36]

Pogroms, and especially the Holocaust, acted as "selection events." More intelligent Jews would have had more resources, negotiated better with gentiles, been better at executing escape and survival plans, and would have, for example, correctly predicted the future and thus have fled Germany in the 1930s and mainland Europe at the beginning of World War II. The result is that Ashkenazi Jews (those from Eastern Europe) have an average IQ of 112, the highest IQ of any ethnic group in the world. Most other Jewish groups are more intelligent than the populations among whom they lived, though the difference is not as

34 S. Cohen, "The Origins of the Matrilineal Principle in Rabbinic Law," *AJS Review,* 10 (1985): 19–53.

35 W. Jacob, "Reform Judaism and Mixed Marriage" in W. Jacob and M. Zemer, (Eds.), *Marriage and Its Obstacles in Jewish Law: Essays and Responses* (Oxford: Berghahn Books, 1999), 222.

36 M. Thomas, M. Weale, A. Jones et al., "Founding Mothers of Jewish Communities: Geographically Separated Jewish Groups Were Independently Founded by Very Few Female Ancestors," *American Journal of Human Genetics,* 70 (2002): 1411–1420.

pronounced as in the case of the Ashkenazi.[37]

Hinduism and the Caste System

The Indian caste system can also be regarded as a eugenic practice, in the sense that it raises the intelligence of certain castes. As we have discussed, ethnocentrism is an important factor in determining which group triumphs over another. However, also important is military technology, battle planning, and organization. These will all be underpinned by intelligence, and in a harsh yet stable ecology intelligence will be duly selected for. This places groups that live in a relatively easy and predictable ecology at a severe disadvantage in the battle of group selection. One strategy which gets round this problem is for the group—through strict endogamy—to create one or many extremely high IQ castes. Males will be naturally inclined to select for youth and beauty and care rather less about the psychological qualities of their spouse, beyond eschewing females that might potentially cuckold them. However, if they are compelled to marry people from an extremely similar socioeconomic background, then they will be compelled—to a great extent—to marry people of similar intelligence. Over a long period of time, this will create very high IQ castes. This would happen if there were selection for higher intelligence within castes, for example when a caste makes its living with some high-skilled occupation such as trading or exegesis of holy texts.

Moreover, if it is the will of gods that you marry within your caste and that you are a dutiful member of your caste, there will be strong social stability. This will reduce the potential for intra-group conflict, meaning that such a system effectively promotes positive ethnocentrism. It can be argued that this is what the Indian caste system helps to achieve, and, to this extent, it is eugenic in nature, or at least socially effective towards group-level eugenic ends. Furthermore, it has been suggested that the system was established to, in effect, keep the "high caste" Aryan invaders—those who were successful in group selection and es-

37 Lynn, *The Chosen People, op. cit.*

tablished the caste system about 1000 BC[38]—separate from the low caste indigenous population.[39] Certainly, there are average caste differences in intelligence, with IQ tending to increase as you ascend the caste system.[40] As one ascends the caste system then the prevalence of European genes becomes greater.[41]

There are four castes—*varna*—in Hinduism, though there is a hierarchy even within these castes, based on the kind of job associated with the caste or the perceived degree of purity associated with the family. These sub-castes are known *jati.* In Hindu myth, the castes descend from different parts of the god Brahma's body. People who are not ethnically "Indian," including all non-Hindus as well as members of tribal groups (whose religion may deviate in important ways from Hinduism) and those who do "impure" jobs, are outside of the caste system. The *Brahmin* (priests and teachers) are the highest *varna*. Beneath these is the *Kshatriya* (warriors) and beneath these is the *Vaishya* (merchants). These are all "twice born," meaning that they are ritually purer than the *Shudra* (servants and labourers) who sit beneath them and must traditionally behave deferentially towards them. Outside of the caste system are the outcastes. These "*Pariahs*" or "*Dalit*" ("broken ones") perform the most unhealthy and degrading tasks and are traditionally regarded as ritually impure, meaning that nobody from a higher caste will breed with them, or even touch them.[42] Within these categories, there are numerous *jati* (also referred to as "castes" by Indians) relating to professions, which were formerly relatively rigidly hereditary, and other factors relating to perceived purity and status. Overall, therefore,

38 R. K. Puthri, *Indian Caste System* (New Delhi: Discovery Publishing House, 2004), 2.

39 See, S. Srivastava, *Passionate Modernity: Sexuality, Class, and Consumption in India* (London: Routledge, 2007).

40 R. Lynn and H. Cheng, "Caste Differences in Intelligence, Education and Earnings in India and Nepal: A Review," *Mankind Quarterly*, 59 (2018): 98–107.

41 M. Bamshad, T. Kivisild, W. S. Watkins et al., "Genetic Evidence on the Origins of Indian Caste Populations," *Genome Research,* 11 (2001): 994–1004.

42 For a readable introduction to Hinduism, see: G. Flood, *An Introduction to Hinduism* (Cambridge: Cambridge University Press, 1996).

there are thousands of castes, usually identifiable by the surname. Currently, in India, only 5 per cent of people marry outside their caste, meaning outside their *jati* rather than their *varna.*[43]

British-Indian journalist Angela Saini (b. 1980) published a book in 2019 bemoaning what she saw as the "return of race science" and of scientists who supposedly espoused "eugenics." I, along with many colleagues, was discussed in the book.[44] The surname "Saini" derives from the lineage having been founded by a son of Maharaja Sainis.[45] Saini is thus a *Rajput* caste, with *Rajput* meaning "son of a king."[46] The *Rajputs* are a caste within the broader *Kshatriya* caste. Angela Saini is, thus, the product of many generations of eugenics. Her husband, journalist Mukul Devichand (b. 1980),[47] is of the same caste as her,[48] so, in effect, she practices eugenics herself while condemning it for others. In 2022, British science journalist Adam Rutherford (b. 1975), who has a doctorate in genetics from University College London, published a book criticising eugenics entitled *Control: The Dark History and Troubling Present of Eugenics.*[49] His mother's surname was Sitaram, part of the high caste *Kalingi*, meaning Rutherford is also the product of many generations of deliberate eugenics.[50] The caste system, it should be noted, is less rigid

43 S. Rajagopalan, "In India, marriages are limited within caste boundaries," *Deccan Herald* (2nd August 2020).

44 A. Saini, *Superior: The Return of Race Science* (Boston, MA: Beacon Press, 2019), 69.

45 Saini Online. (2018). "The origin of Sainis," http://www.sainionline.com/origin-of-saini-caste-from-yadu-rajputs.

46 See J. Asopa, *Origin of the Rajputs* (Varanasi: Bharatiya Publishing House, 1976).

47 Keble College, Oxford, *The Record* (2009), 132.

48 M. K. Ghilla, *The Guru Consorts* (Delhi: Radha Publications, 1992), 50.

49 A. Rutherford, *Control: The Dark History and Troubling Present of Eugenics* (London: Orion, 2022).

50 See, J. H. Hutton, *Caste in India: Its Nature, Function, and Origins* (Oxford: Oxford University Press, 1969); J. Frazer, *Totemism and Exogamy: A Treatise on Certain Early Forms of Superstition and Society* (London: Macmillan, 1935), 231.

and less delineated in southern India.[51] The Kalingi caste is from southern India, specifically from the states of Andhra Pradesh and Odisha.

At the extremes, the highest caste *Brahmin* are supposed to follow what is effectively a very healthy vegetarian or even vegan diet, but some very high castes take this even further. The *Iyengar Brahmin* eschew root vegetables, including garlic and onion.[52] By contrast, the *Dalit* can eat whatever they like and, indeed, they do eat such animals as rats and snakes, with some *Dalit* castes simply hunting these for a living. Accordingly, it can be argued that the Hindu dietary laws ensure that those who, on average, are likely to be higher in intelligence follow a healthier diet, although a vegan Brahmin is more likely than a rat-eating Dalit to suffer from deficiencies of protein, zinc, vitamin B12, and iron.[53] Overall, though, it might be averred that this dietary system is likely to mean that the higher castes will live longer and so enjoy higher completed fertility. By contrast, members of lower castes are permitted to eat a less healthy diet, which we might actually expect to be more attractive due to its inclusion of meat. So, in a small way, the Hindu dietary laws could be seen to push the population in the direction of higher intelligence. By ensuring this, the group is more likely to triumph in the battle of group selection. However, any group which made these kinds of dietary laws a religious issue would be at a distinct advantage because the rules would be more likely to be obeyed. Thus, on a certain level, dietary laws in religious groups are a matter of group selection, which will ultimately help to push the nature of the group in a certain direction. It should also be noted that there exist caste systems in many Sub-Saharan African countries under which there are hereditary occupations and notions of certain castes

51 "Tamil" in J. Minahan, (Ed.), *Encyclopedia of Stateless Nations* (Westport, CT: Greenwood Press, 2002).

52 A. Dhillon, "In India, caste system ensures you are what you eat," *Post Magazine* (26th July 2014), http://www.scmp.com/magazines/post-magazine/article/1558061/you-are-what-you-eat.

53 On vegan health deficiencies, see G. Rizzo, A. Lagana, A. Rapisarda et al., "Vitamin B12 among Vegetarians: Status, Assessment and Supplementation," *Nutrients*, 8 (2016): 767.

being impure.[54]

The Christian World

Despite the monogamy of the Christian world, there is evidence that selection for wealth continued in the form of *de facto* polygamy. The nobility would generally have mistresses in addition to their wives—often servant girls working in their households. Until the Reformation, it was so socially acceptable for the upper class to have illegitimate children by mistresses that these "noble bastards" or "royal bastards" would be acknowledged by their father, take their father's surname, be raised in his household and be provided for by him, often lavishly.[55] English historian Stephen Staves has claimed that upper class men in Early Modern England would generally have "roughly as many illegitimate children as legitimate ones."[56]

Even among the supposedly celibate clergy, there were many cases of priests fathering children. Pope Alexander VI (1431–1503) had four of them, and Lucrezia Borgia (1480–1519) was his illegitimate daughter.[57] It should be emphasised that priests were found well-down the social hierarchy, in the form of "poor priests" who combined their calling with being farm labourers and many had little education,[58] so it is unlikely that priestly celibacy heavily interfered with more intelligent males tending to be highly fertile. Thus, there is evidence that even in the pre-Modern, Christian world a form of *de facto* polygamy continued such that the upper class could have pronounced fertility compared to

54 C. Chucks Osuji, *Inequality, Tension and Conflict* (Bloomington, IN: XLibris, 2018).

55 K. Carlton and T. Thornton, "Illegitimacy and Authority in the North of England, 1450–1640," *Northern History,* XLVIII (2011): I.

56 S. Staves, "Daughters and younger sons" in J. Brewer and S. Staves, (Eds.), *Early Modern Conceptions of Property* (London: Routledge, 2014), 210.

57 S. Bradford, *Lucrezia Borgia: Life, Love and Death in Renaissance Italy* (London: Penguin, 2005).

58 See, E. Dutton, *The Ruler of Cheshire: Sir Piers Dutton, Tudor Gangland and the Violent Politics of the Palatine* (Northwich, Cheshire: Leonie Press, 2015).

the lower classes. However, it is as we move into the Early Modern Era—with increased record keeping—that we can find direct evidence that the fertility of the rich, even though they were only a small percentage of the population, was much higher than that of the poor. Polygyny was most likely a minor contributor. The "Survival of the Richest," a term coined by Scottish economist Gregory Clark,[59] was mainly because borderline nutrition and hard physical work reduce female fertility, which is why rich women had shorter inter-birth intervals.

Fertility in the Early Modern Era

From around the beginning of the seventeenth century, most English parishes began to keep systematic parish records of baptisms, marriages, and burials. Some parishes, right from the beginning, kept very detailed records, including the names of the child's parents on the baptism record and the name of the father in the case of an infant burial. The survival of wills, proved by parish courts, is sketchy in the sixteenth century, but is relatively comprehensive by the seventeenth century. As such, we now have a series of fascinating studies that give us real proof that the rich were outbreeding the poor in Early Modern England and elsewhere in Europe.

The most detailed of these studies was produced by Gregory Clark, whom we met earlier. He shows, drawing upon 1,978 wills from Suffolk and Essex, in the east of England, made between the years 1585 and 1638, that the richer 50 per cent of English testators had many more surviving children than the poorer 50 per cent. The completed fertility of the richer half was 40 per cent higher than that of the poorer half. Wills extended well down the social hierarchy.[60] In 1978, historian Victor Skipp (1925–2010) found, from a sample of Warwickshire parish records between 1560 and 1670, that the "middle class" (the richer half) had, on average, four children while the "working class" (the poorer half)

59 G. Clark, *A Farewell to Alms: A Brief Economic History of the World* (Princeton, NJ: Princeton University Press, 2007).

60 Clark, *A Farewell to Alms*, *op. cit.*, 87.

had three.[61] This was "children baptised" rather than "completed fertility," meaning that, as 45 per cent of children tended to die in infancy, some of the poorest may have ended-up with no surviving children at all.

In fact, in 1972, local historian John Pound found, drawing upon Norfolk records, that between 1500 and 1630 completed fertility was four for the middle class and two for the working class, giving the middle class a fertility advantage of 100 per cent.[62] A similar pattern has been unearthed in all pre-industrial European societies. For example, according to research by German historian and psychologist Volkmar Weiss, in Saxony, between 1547 and 1671, the middle class, on average, had 3.4 children who got married while the working class had 1.6 who did so.[63] So, the seventeenth century continued to see the pattern which we see even in pastoralist tribes. The wealthier half of the population were simply more fertile—they left more surviving children—than the poorer half of society, and there were further gradations in favour of the richer even within the "rich" category. In addition, the rich had a far longer life expectancy than the poor. For example, data from seventeenth century Geneva shows that the upper class lived, on average, until the age of 35.9, the middle class until 24.7, and the working class until 18.3.[64]

Here one consideration is that attainment of wealth and social status required different traits in complex Early Modern societies than in primitive tribes. A Yanomamö chief could attain his position and have multiple wives by killing people in "war" (actually, ambushes near neighbouring villages). Physical strength and agility, fearlessness and risk-taking were most likely more important than intelligence for this, though intelligence would still have been relevant. But in complex societies people got rich

61 V. Skipp, *Crisis and Development: An Ecological Case Study of the Forest of Arden 1570 1674* (Cambridge: Cambridge University Press, 1978).

62 J. F. Pound, "An Elizabethan census of the poor," *University of Birmingham Historical Journal*, 7 (1972): 142160.

63 V. Weiss, "Social and demographic origins of the European proletariat," *Mankind Quarterly*, 31 (1990): 126152.

64 Lynn, *Dysgenics, op. cit.*

with trade and in other occupations that required literacy and intelligence. Therefore, it may be that for some time there was divergent selection, with stronger selection for high intelligence in complex societies with many different economic niches requiring complex skills than in primitive societies.

The high child mortality rate, of around 50 per cent, meant that the population grew only very slowly. And when it grew too high—higher than the capacity of the land to be able to sustain it—then there would be widespread famine and the population would dramatically decline. The consequence of this was that England, and other European societies, were characterized by a system of constant social descent. Every generation, those at the bottom of the hierarchy would die off, without children, and those who were one step up the ladder would, by necessity, move downwards to take their place. In general, the younger sons of the gentry would fall into the "middling sort." They would become yeoman farmers who, on average, weren't as wealthy as the gentry and would do some farm labour themselves.[65] Or these younger gentry might become merchants, but, either way, they would work for a living. The younger sons of merchants would become craftsmen and the younger sons of yeomen would be husbandmen, meaning they would take to the plough themselves. The younger sons of craftsmen and husbandmen would be cottagers, who would supplement a smallholding with day labour on the farms of others; and the younger sons of smallholders would simply be labourers, who made up the majority of the population. And below these were the destitute and the starving. As such, we would expect that the qualities that made people socioeconomically successful—including highly genetic qualities such as intelligence—would be growing in the population every generation. The genes for being rich—which is significantly predicted by intelligence, and which is strongly heritable—were being selected for under the harshness of pre-industrial conditions. We would predict that people would have been becoming more intelligent; the average IQ of pre-industrial society would have increased every generation.

65 For further discussion of the nature of these social ranks, see: Dutton, *The Ruler of Cheshire, op. cit.* Ch. 2.

Upward Social Mobility

We know that in pre-industrial England, the richer half of the population had higher fertility than the poorer half and there is evidence that intelligence was the reason for this, even then. This can be better understood if we look at the nature of social mobility in Medieval and Early Modern England. Many people believe that social mobility is a modern phenomenon and that your childhood social status pretty much dictated your life further back in history, but this is simply untrue. Gregory Clark has charted the rise and fall of particular families over time, by focusing on those with unusual surnames. In his book *The Son Also Rises*, Clark presents data for assorted countries including England, Japan, the US, China, Sweden, and India. Clark notes that social mobility, in modern times, is often assumed to be high because there is a weak correlation, of around 0.3, between the income of parent and child. But this is problematic because there are different measures of status: wealth, education, and occupational status, for example. People can make trade-offs between these measures. A teacher would have higher occupational status than a plumber, but may well earn less. In addition, argues Clark, because the factors which affect social status are likely to be strongly genetic, there will be random fluctuation when comparing father and son. As such, it is better to compare surnames—and thus families—across time.[66]

In England, for example, Clark argues that you can tell the historical social status of a family by the kind of surname they have. High status surnames are those which are Norman (such as those that end in "ville") or those which are "locative"—the names of places. People with these surnames are descended from Normans, who took the name of their feudal manor. Surnames which refer to a profession—Bailey, Cooper, Thatcher—are middle ranking, while low ranking surnames tend to end in "son," to be the name of the father (e.g. "John"), refer to physical appearance (e.g. "Brown"), or relate to the part of the village a person lived in, such as "Hill." Assessing the data between the Medieval

66 G. Clark, *The Son Also Rises: Surnames and the History of Social Mobility* (Princeton, NJ: Princeton University Press, 2014).

period and 2012, Clark finds that across the period, and even now, those with Norman or locative names are over-represented among proxies for high social status. These include Oxbridge graduates, barristers, and physicians. Those with low status surnames are under-represented. Over this period, Clark emphasises, there has been a gradual "regression to the mean." This means that those at the representation extremes have moved closer to the average over time, so that Normans are less over-represented than they used to be, and Saxons are less under-represented than they used to be.

Clark provides a number of explanations for this. Something akin to regression to the mean happens in genetics. Owing to the large number of genes involved, it sometimes (though rarely) happens that children are significantly more (or less) "socially competent" (Clark's term for the ability to achieve high socioeconomic status) than their parents. They then marry a person with comparable abilities and move up (or down) the social hierarchy. Alternatively, a person with high "genotypic ability" marries a person with high "phenotypic ability." Finally, as pre-modern fertility was predicted by high social status, Norman surnames would have to spread downwards through the population.

Overall, Clark finds that social status across history is around 0.75 heritable, its heritability is the same now as it was in Medieval England, and sudden shocks that might raise social status very quickly—such as winning the Lottery—wear off within a few generations. This implies that ability is required to maintain wealth and a person of low ability who is born into a wealthy family will likely squander their inheritance. This movement towards the mean, in terms of surnames and status, implies that something is being selected for in all classes, and this is likely intelligence, though Clark doesn't look directly at this trait. Interestingly, Clark finds the same results in countries that are far more egalitarian than Britain, such as Sweden. Dividing between the surnames of the higher nobility, the surnames of the untitled nobility, Latin surnames (historically adopted by the highly educated non-noble), geographical surnames (e.g. Berg) and the lowest status surnames (those ending in "son"), Clark finds a similar pattern to England. Those with noble and Latin surnames are still over-represented at the top universities, amongst the richest, and in the top professions, and the heritability of social status is about 0.75.

Illegitimates

So, in pre-industrial England, wealth predicted fertility, and the way you obtained or maintained wealth was through high intelligence. At the other end of the social scale there is evidence that those who had particularly low intelligence would be especially likely to *not* pass on their genes. In general, illegitimacy—especially in the form of single motherhood—is associated with low status and, so, low intelligence. In the US, White American single mothers have an average IQ of 92, whereas it is 105 for women who are childless or married with children. We have seen that education level is a proxy for intelligence. Research from the US has found that women with no high school education are 20 times more likely to end up as single mothers than are women with a high school education.[67]

In pre-industrial Europe, there was no welfare state. This generally didn't develop until the second half of the twentieth century. As such, single mothers would tend to abandon unwanted babies. In the early nineteenth century, in London, it was not uncommon to see dead babies in the streets or in rubbish dumps. By the eighteenth century, the number of abandoned babies was so great in many European cities that orphanages were established to house them. In 1741, the Thomas Coram Hospital for Foundlings was opened in London. However, due to the lack of wet nurses, 71 per cent of these foundlings were dead by the age of 15, whereas it was roughly 40 per cent in the general population. Due to insufficient wet nurses, foundlings were malnourished and acutely vulnerable to infectious disease.[68] Thus, in pre-modern societies, illegitimate children of single mothers, who would have been of low intelligence, suffered very high mortality. Each generation, this would have acted to stop genes

67 R. Herrnstein and C. Murray, *The Bell Curve: Intelligence and the Class Structure in American Life* (New York: Free Press, 1994).

68 D. Coleman and J. Salt, *The British Population* (Oxford: Oxford University Press, 1992).

for low intelligence spreading through the population.

Execution

Canadian anthropologist Peter Frost and American anthropologist Henry Harpending (1944–2016) have examined the way in which judicial violence acted as a selection pressure in pre-industrial Europe.[69] Until the eleventh century, they showed, execution was not widely employed in Europe. This was because the Church was opposed to it, law enforcement was rudimentary, and it was believed that people should have the right to settle their own disputes, such as through duels. But as the Medieval Era progressed, the Church accepted that the "wicked" should be executed so that the "good" could live in peace. By the Early Modern Era, all felonies carried the death penalty as, once more, this was religiously justified, drawing upon the dictates of the Old Testament: "But if there is any further injury, then you shall appoint as a penalty life for life" (Exodus 21:23). In other words, those who behave in a violent manner are to be discouraged from doing so through agonizing punishment and mutilation while those who take another life are themselves to be killed.

Frost and Harpending calculated that this meant that up to 1 per cent of the male population of Europe was executed each generation, with approximately another 1 per cent dying at the scene of the crime or in prison while awaiting trial. Most of these felons, Frost and Harpending demonstrate, were young men. As they were young men, they would have had their fertility curtailed by virtue of their premature deaths. Frost and Harpending argue that this process would have altered the nature of Western personality, by preventing those with high psychopathic personality from passing on their genes. They use something called the Breeder's Equation to calculate whether the prevalence of execution would have had the proposed effect, and they find that it would have. But their model also has implications for intelligence. Those who were executed or died in prison were over-

69 P. Frost and H. Harpending, "Western Europe, State Formation, and Genetic Pacification," *Evolutionary Psychology*, 13 (2015): DOI: https://doi.org/10.1177/147470491501300114.

whelmingly poor and uneducated. In England, those who were of high social status could fund relatively luxurious conditions in prison and, unless their crime was treason or heresy, they could avoid execution by pleading "Benefit of the Clergy." This meant if they could read then they would avoid execution. It also meant that it was disproportionately those of low intelligence who were executed, as those of high intelligence would be, in general, able to read and so plead Benefit of the Clergy. In addition, we have already seen that overall, intelligence is negatively associated with criminality, and we might expect this association to be particularly strong when the punishment for a felony is death. This would mean that judicial execution was very likely playing a role in boosting European intelligence by removing some of the least intelligent young men every generation.

Swedish psychologist Guy Madison and I have developed this theory.[70] We argued that the widespread use of execution in Medieval England would have increased levels of religiousness. We compared genetic correlation data for traits predicting religiousness, the heritability of religiousness, and the evidence for increasing religiousness across the Middle Ages. We found that levels of religiousness significantly increased between 1400 and 1500, as evidenced by a large rise in the per capita numbers of people joining religious orders across the period. Madison and I found that the high level of execution would have played a role in selecting for increased religiousness in England. We also explored the mechanisms by which it would have done so. Madison and I show that religiousness is associated with high altruism and Conscientiousness and low levels of autism (autism is characterized by an inability to empathize, which also causes low religiosity, for reasons we will explore later), low levels of suffering from stress, and with a feeling of being watched. All of these crucial dimensions to religiousness are negatively associated with rule-breaking and criminality, we demonstrated. So, in Medieval England there was an extent to which the government was, through its policy of widespread execution, preventing the

70 E. Dutton and G. Madison, "Execution, Violent Punishment and Selection for Religiousness in Medieval England," *Evolutionary Psychological Science*, 4 (2018): 83–89.

relatively irreligious from passing on their genes. It was acting in such a way as to make the society more religious and thus even more law-abiding.

Is there Evidence that Intelligence Was Increasing Up to the Industrial Revolution?

Clearly, there were no IQ tests in pre-industrial England allowing us to track the rise of IQ. Gregory Clark, though he does not explicitly talk about the relevance of intelligence, has amassed very good evidence that intelligence was nonetheless continuously increasing in the pre-industrial era. He has done this by gathering proxies for intelligence and showing that, across history, they seem to change in precisely the direction that would be predicted if intelligence was indeed going up. These are interest rates, literacy, and numeracy.

Interest Rates

Interest rates can be regarded as a marker of intelligence because they measure time preference. We have already seen that "time preference" is associated with intelligence. More intelligent people are more focused on the future than are less intelligent people. As such, a smaller reward, given relatively further into the future, is sufficient to persuade a more intelligent person to postpone immediate gratification. It follows from this that if you asked a less intelligent person to lend you some money, they would be profoundly concerned about the immediate consequence of this, which would be them having slightly less money here and now. In order to persuade them to lend you the money, you would have to offset this by allowing them to charge you a very high level of interest. If they could only charge you a low level of interest then, from their perspective, the reward of the low level of interest would be too small, relative to the length of time they would have to wait for it and so be ungratified, to matter. It would be outweighed by the fact that they would be postponing gratification here and now. In essence, the more intelligent a person is, the less of a return they require on their investment to bother making the investment.

In addition, the intelligence of the debtor is highly relevant to the fall in interest rates across time. Less intelligent people cannot fully appreciate that they have to pay back the money that they borrow and that they may not be able to do so. Therefore, they are prone to borrow more than they can pay back, leading to a high default rate. This is what happened when the mortgage market in America went bust in 2008, and the lenders recognized too late that many of their low-IQ clients couldn't pay back their mortgages because they had miscalculated their financial muscle.[71] Today we have a jungle of regulations to deal with this problem and governments who rescue the banks that are "too big to fail," but in earlier times the lenders had to demand high interest rates so that the successful loans could compensate for the substantial losses they suffered from people who never paid back the borrowed money.

Low interest rates, then, are a marker of societal intelligence. Clark's research has shown that between 1200 and 1800 in England, interest rates significantly fell. Based on land return and rent return, interest rates in the year 1200 were over 10 per cent and, in fact, in 1150 they were around 15 per cent. By 1800, they had fallen to just 5 per cent. Of course, there are all kinds of localized reasons why interest rates might have fallen—and fallen so much—over this period. My point is simply that it fits with the hypothesis that people were becoming more intelligent—and thus better able to defer gratification.

Literacy and Numeracy

Clearly, literacy and numeracy are related to intelligence. "Verbal intelligence" is an important component of intelligence and thus, the more intelligent you are the easier you will find it to learn to read. Intelligence correlates with "intellect"—being interested in ideas. The more intelligent you are, the higher your intellect will be and thus the more of an incentive you will have to learn to read. As for learning to write, there will be clear benefits to

71 For an interesting discussion of this crisis, see the memoirs of the British Chancellor of the Exchequer at the time: A. Darling, *Back from the Brink: 1000 Days at Number 11* (London: Atlantic Books, 2012).

learning to do this in terms of keeping records and accounts, and these will become more necessary as society becomes more complex and specialised, which would potentially be a sign of its increasing intelligence. In addition, writing—especially with a quill—is a complex skill involving fine motor skills and patience, both qualities associated with more intelligent people. Furthermore, the wealthy would be more likely to be able to educate their children, and those children would be more able to accrue wealth by being educated. It is, therefore, no surprise that Clark found a relationship between literacy and social status in his sample of wills. He found that 94 per cent of the gentry could sign their wills, 88 per cent of merchants, 53 per cent of yeomen, 26 per cent of husbandmen, but only 17 per cent of labourers. So, we would expect literacy to be a proxy for intelligence.

The same would be true of numeracy. More intelligent people would be better at mathematics and the kind of work they engaged in, such as business, would also necessitate the greater use of it. Societies that are undeveloped have little use for mathematics. Clark notes that a good marker of numeracy is age awareness. By 1800, most people in Western Europe had a fairly accurate understanding of how old they were. However, this is not the case in illiterate and innumerate societies. In the fourth century AD, a survey indicated that 80 per cent of Roman office holders—highly important people—knew their age. In eighteenth century Paris, only 15 per cent of people were of unknown age. In fifteenth century Florence, a wealthy city, 32 per cent of people did not know how old they were. In the small town of Corfe Castle in England, most of whose inhabitants were labourers, only 8 per cent did not know their age in 1790. This implies that English labourers around 1800 were as numerate as late Roman office holders.

For literacy, Clark turns to a combination of the percentage of grooms who signed the marriage register when they married or the percentage of witnesses in court cases who could sign depositions. These were traced between 1580 and 1920 in England, and showed a clear upward trend. For example, in 1740 about 60 per cent of English men could sign their names but by 1920 it was close to 100 per cent, gradually increasing, without a blip, from 1780. A similar process can be seen with female literacy and literacy from specific bishoprics in narrower periods. Again, factors such as the English Civil War, for example, would interfere with the trend—because they would interfere with the education of children—but there is a clear upward trend. There is no sign that there were greater rewards for literacy and numeracy in 1800 than in 1500; we simply became more educated. This implies we became more intelligent, either because higher intelligence promoted education or because education raised intelligence. Most likely both causal arrows were at work.[72] In addition to the measures discussed by Gregory Clark, there are a number of others which fit with the pattern he observed.

The Size of the Head

Skull size and head circumference correlate about 0.2 with IQ, a relationship that is further evidence of the objectivity of IQ tests.[73] On average, the larger your brain is, the more intelligent you are. Large brains require large skulls and, this being so, we would expect that the average size of people's heads would have increased since, for example, the Medieval period. Comparing a sample of 31 skulls from thirteenth century London and 30 skulls that are representative of modern English people, a study found that the cranial vault, which houses the brain, had increased in size by 10mm.[74]

72 Clark, *A Farewell to Alms, op. cit.*

73 J. Pietschnig, L. Penke, J. Wicherts et al., "Meta-analysis of associations between human brain volume and intelligence differences: How strong are they and what do they mean?" *Neuroscience and Biobehavioral Reviews,* 57 (2015): 411–432.

74 W. P. Rock, A. M. Sabieha and R. Evans, "A cephalometric comparison of skulls from the fourteenth, sixteenth and twentieth centuries," *British Dental Journal,* 200 (2006): 33–37. It should be noted that the research team had originally thought that the skeletons were from the mid-fourteenth century. It was later realized that they were from the thirteenth century. See: R. G. Steen, *Human Intelligence and Medical Illness: Assessing the Flynn Effect* (New York: Springer, 2009).

Per Capita Major Innovation

Yet another measure is per capita major innovation. When intelligence is rising, then the "smart fraction" of the population—the geniuses, who have extremely high intelligence—will become smarter and more numerous. This will result in an increase in per capita major innovations. Consistent with this, the number of per capita major inventions per year increased continuously from the Middle Ages until around 1870. The same occurred with the use of high order words (abstract words) in published texts. Our heads grew bigger and bigger to accommodate our growing brains, literacy and numeracy increased despite little change in living standards, and alleles associated with high intelligence became increasingly common in European populations.[75] Thus, it is clear that intelligence was increasing across time, and that this is what set off the Industrial Revolution. How did this lead to intelligence going into decline?

75 Dutton and Woodley of Menie, *At Our Wits' End*, *op. cit.*, Ch. 4.

Four

The Fall of Intelligence

The Collapse of Intelligence in Miami

In the early hours of 24th June 2021, Champlain Towers South—an exclusive, twelve-storey condominium overlooking the beach in the Miami suburb of Surfside—partially collapsed. Thirty-five people were rescued from the part of the building that remained standing. Needless to say, those who lived in the part that fell were rather less fortunate. By mid-July, 94 of its occupants had been confirmed dead, 11 were injured, and 22 were unaccounted for. One person was rescued from the resultant rubble.[1] The fall of Champlain Towers South—known as the "Surfside Building Collapse"—was one of the deadliest building collapses in American history. Devastating as the loss of life was, it was rendered all the more poignant by the fact that it could so easily have been avoided if the people who ran the property had been slightly more intelligent.

In 2018, a report by structural engineer Frank Morabito—an individual whom we would expect to have very high intelligence, as we will see shortly—made it perfectly clear, at least to people of above a certain level of intelligence, that Champlain Towers South was definitely going to collapse if nothing was done to

1 R. Ellis, H. Silverman, M. Alonso and K. Jones, "Death toll in Surfside condo building collapse reaches 94," *CNN* (13th July 2021), https://edition.cnn.com/2021/07/12/us/miami-dade-building-collapse-monday/index.html.

rectify what was going on. The engineer explained in his report: "The failed waterproofing is causing major structural damage to the concrete structural slab below these areas [...] Failure to replace the waterproofing in the near future will cause the extent of the concrete deterioration to expand exponentially." He added that these changes needed to occur "in a timely fashion."[2] However, despite the engineer explaining that the deterioration would increase "exponentially" (at an accelerating pace) and that this needed to be dealt with "in the near future" and "in a timely fashion" (in other words "soon"), three years later, nothing had been done. The simplest explanation for this failure is a relative lack of intelligence on the part of those who were tasked with managing the building and even, perhaps, on the part of those who were tasked with managing the city of Miami itself.

Low intelligence predicts that you will fail to think about the future, preferring to live in the present. This "time preference" means that you will be more likely to prefer an easier situation in the present, such as saving money by not carrying out vital building repairs, than opt for a difficulty in the present—spending money—in pursuit of a preferable future: avoiding the collapse of a building. The less intelligent people are, then the less oriented they are towards the future. In the well-known marshmallow test, children are given the option of being given one marshmallow now or two marshmallows in five minutes' time. More intelligent children almost always wait for the greater reward; less intelligent children cannot do this.[3]

Low intelligence also predicts having a poor vocabulary. This is because words are "thinking tools," meaning that you can solve more complex problems better with greater and more subtle thinking tools at your disposal. The engineer did not spell out in everyday language what was going to happen and, thus, less intelligent people would simply not have understood the gravity of the situation. To a more intelligent person, however, Morabito was perfectly clear. The BBC even stated: "His report didn't

2 *BBC News*, "Florida building collapse: Report from 2018 warned of 'major damage,'" https://www.bbc.com/news/world-us-canada-57621774.

3 S. Pliszka, *Neuroscience for the Mental Health Clinician* (New York: Guildford Publications, 2016), 169.

suggest the 40-year-old building was at any imminent risk of collapse, but he urged that the concrete repairs be carried out in 'a timely fashion.'" This is crucial. A low IQ person would have acted to save the building if Morabito had written "make these changes or the building will collapse very soon." He didn't write this, because the collapse was not "imminent", and he assumed it should be obvious to people that "exponential" damage to the building's structural integrity strongly implied that it would eventually collapse.

Frank Morabito assumed that his report would be read by people who were as intelligent as him, or at least not very significantly less intelligent than him. Unfortunately for the almost 100 people who died and for their relatives, Morabito was wrong about this. The current generation and the kind of people who manage buildings now are significantly less intelligent than when the buildings were erected, meaning that we increasingly live in an over-promoted society compared even to the 1970s, as we will discuss in more detail below. How has this situation come about?

Why Did Intelligence Go Into Decline?

The Industrial Revolution, as noted above, changed the system of harsh Darwinian selection to which we had previously been subject. By permitting the manufacture of cheaper clothing, it made countries wealthier, as the production of resources increasingly outpaced the rise in population, leading to much greater abundance in all manner of materials via increased mechanization. This wealth could be invested in important projects, the projects could be pursued more cheaply, and, due to there being a very smart "smart fraction," all manner of people were making all kinds of important breakthroughs in all kinds of areas anyway. However, greater wealth created more people with more free time and disposable real income, setting them free to come up with more genius innovations or micro-innovations, in a positive feedback loop. Conversely this means that as standards of living fall, and as the smart fraction and general population become less intelligent, the sum total of people have less time and resources to address the problems causing the fall in living standards in the first place. Thus, they are less able to "invent their way out of it," even if they had the ability to do so, being too busy foraging for food and guarding their property from violent mobs. This is the negative feedback loop which will eventually develop in a society of declining intelligence.

However, the late eighteenth century positive feedback loop, set off by inventors such as Richard Arkwright, led to, for example, improved medicine and better public health. This meant that child mortality—which was a correlate of low intelligence, using the proxy of wealth—collapsed. In 1800, child mortality was approximately 50 per cent, though it was much lower among the more intelligent (the wealthier) and much higher among the less intelligent (the poorer). By 1900, the infant mortality rate in England wasn't 50 per cent, but a mere 10 per cent.[4] Almost every child survived into adulthood and the relationship between poverty and child mortality would have been very weak. Hence, by 1900, there was no longer a positive relationship between wealth (and thus IQ) and completed fertility. Mutation was being selected out to a much lesser extent, which meant that intelligence—which, as we have seen, is part of a general fitness factor—was being selected for less intensely, and low intelligence was being selected against less intensely. The result was a huge rise in descendants for the lower classes of 1800—those with the lowest IQs.

Fundamentally this change in the social dynamics of successful breeding was caused by the "demographic transition." This is when a society moves from being one characterized by high fertility and high child mortality to being one characterized by low fertility and low child mortality. This transition, which was complete in Western countries by the beginning of the twentieth century, comes in two stages. In the first stage there is a massive decline in child mortality, and this is followed, some years later, by a steep drop in fertility. In the intervening period, the population will increase dramatically because there will be *both* high fertility and low child mortality, presumably because people have not yet quite grasped that it is safe to heavily limit their fertility, thereby allowing their living standards to improve, and still have

4 E. Hawe, *Compendium of Health Statistics, 2009* (Manchester: Radcliffe Medical, 2009).

the certainty of having at least one or two surviving children. By the middle of the twentieth century, most families in Western countries had only 2 children and, by the end of it, they were down to fewer than 2—to below replacement fertility. Some demographers have defined a "second demographic transition," the one that brought fertility below replacement in the second half of the twentieth century. This is the one caused by gender equality, career women, and child-free lifestyles.[5]

Contraception

A number of factors created a negative relationship between intelligence and fertility. Reliable contraception was developed, the successful use of which is associated with impulse control, forward-planning, and thus with intelligence. This allowed the more intelligent who desired fewer children than the less intelligent anyway,[6] for reasons we will explore later, to have fewer children, meaning that large families became a function of low intelligence. Higher in Openness-Intellect than the less intelligent, the more intelligent were also, for this reason, less prone to instinctive ideals, such as desiring one's own children, an issue to which we will return later when we explore the evolution of personality.

It should be noted that, contrary to popular belief, drawing on Polish anthropologist Bronislaw Malinowski's (1884–1942) fieldwork with the Trobriand Islanders, there is no known human society that has not understood the relationship between sex and pregnancy, presumably having observed the relationship from animals.[7] Thus, it is perfectly possible for humans to have developed an instinctive desire to have children, as well as to have sex and to care for children. Indeed, there is evidence

5 For discussion, see: B. Zaidi and S. P. Morgan, "The Second Demographic Transition: A Review and Appraisal," *Annual Review of Sociology*, 43 (2017): 473–492.

6 S. Kanazawa, *The Intelligence Paradox: Why the Intelligent Choice Isn't Always the Smart One* (Hoboken, NJ: John Wiley & Sons, 2012), 179–180.

7 See: B. Senft and G. Senft, *Growing Up on the Trobriand Islands in Papua New Guinea: Childhood and Educational Ideologies in Tauwema* (Amsterdam: John Benjamins Publishing, 2018), 17

that females have "baby fever," where they specifically wish to be pregnant, and that this is partly heritable.[8] Further, it has been found that priming people with mortality salience—which will tend to make people stressed and so in touch with their instincts and cognitive biases—increases their desire to have children.[9]

Feminism

The rise of feminism meant that more intelligent women would dedicate their most fertile years to their higher education and careers, meaning they would only have a small number of children, with the most intelligent women having no children at all. By contrast, females with low intelligence would start having children at a young age, have many children with a variety of men, and they would become grandmothers by the time their more intelligent contemporaries were becoming mothers. Thus, low IQ women don't just have more children: they also have more generations. Leading English eugenicist, and Eugenics Society vice-chairman, Havelock Ellis (1859–1939), who coined the term "social hygiene"[10] in a book which included the phrase in the title,[11] argued that feminism could be eugenic because liberated females would only choose genetically healthy men, never being forced to procreate with sub-optimal males.[12] But it seems that other factors have overwhelmed this and rendered feminism dysgenic.

8 A. Rotkirch, "All that she wants is a (nother) baby'? Longing for children as a fertility incentive of growing importance," *Journal of Evolutionary Psychology,* 5 (2007): 89–104.

9 I. Fritsche, E. Jonas, Peter Fischer et al., "Mortality Salience and the Desire for Offspring," *Journal of Experimental Social Psychology,* 43 (2007): 753–762.

10 A. Taylor Allen, "Feminism and Eugenics in Germany and Britain, 1900–1940: A Comparative Perspective," *German Studies Review,* 23 (2000): 477–505.

11 H. Ellis, *The Task of Social Hygiene* (New York: Houghton, Mifflin and Co., 1912).

12 I. Crozier, "Havelock Ellis, eugenicist," *Studies in History and Philosophy of Science Part C: Studies in History and Philosophy of Biological and Biomedical Sciences.* 39 (2008): 187–194.

There also appears to be a "Trickle Effect" once high-status females begin limiting their fertility, as people tend to imitate those of higher status.[13] Evidence from Brazil indicates that the rise of soap operas—portraying wealthy women with small numbers of children—may have caused poorer women to limit their fertility,[14] leaving only the most impulsive having large numbers of offspring.

Welfare

The rise in the welfare state also meant that there was no longer an incentive for those low in intelligence to limit their fertility. Child support payments meant that there was actually a positive incentive for them to have children.[15] At the time of writing, in the UK only families where both parents are on welfare and are concomitantly recognized "problem families," requiring legal and social interventions, which implies an IQ of around 80, are breeding at above replacement fertility. Families in which both parents work (IQ about 100) have fewer children than those where one parent is on welfare and they in turn have fewer children than families where both parents are on welfare.[16]

Volkmar Weiss, whom we met earlier, has implied that the spread of democracy may also impact the relationship between fertility and intelligence, insomuch as each vote is of equal value and, to the extent that those of low socioeconomic status vote, they will vote for parties that redistribute resources from the productive to them, allowing them to be ever less concerned about limiting their fertility.[17]

13 G. Simmel, "Fashion," *American Journal of Sociology*, 62 (1957): 541–558.

14 E. La Ferrera, "Soap Operas and Fertility: Evidence from Brazil," *American Economic Journal*, 4 (2012): 1–31.

15 E. Dutton and M. A. Woodley of Menie, *At Our Wits' End: Why We're Becoming Less Intelligent and What It Means for the Future* (Exeter: Imprint Academic, 2018), Ch. 7.

16 A. Perkins, *The Welfare Trait: How State Benefits Affect Personality* (London: Palgrave Macmillan, 2016).

17 V. Weiss, "The Population Cycle Drives Human History: From a Eugenic Phase into a Dysgenic Phase and Eventual Collapse," *Journal of*

IQ and Fertility

The result of all this is that in developed countries there is a negative correlation of about -0.1 between IQ and fertility. Among females it is around -0.2 while among males it is -0.089, attaining statistical significance.[18] Even controlling for education and earnings, IQ among Western females remains negatively correlated with fertility.[19] This negative IQ-fertility nexus is a "Jensen Effect": it is on "general intelligence," which is strongly genetic.[20] The male-female difference likely exists, in part, because males do not have to make a trade-off, in the same way, between their fertility and their career and, of course, males remain fertile for longer. This negative relationship is increasingly being found in developing countries, such as China, as they continue to industrialize.[21]

Interestingly, highly egalitarian and developed societies, such as those in Scandinavia, appear to still have a very weak positive association between fertility and intelligence or between fertility and education level, at least among males, at least according to some studies.[22] In Sweden, among males, there is a positive IQ-fertility gradient based on army conscription data, partly due to very low IQ males having no children at all.[23] However, these

Social, Political and Economic Studies, 32 (2007): 327–358.

18 G. Meisenberg, "The reproduction of intelligence," *Intelligence,* 38 (2010): 220–230.

19 S. Kanazawa, "Intelligence and Childlessness," *Social Science Research,* 48 (2014): 157–170.

20 M. A. Woodley and G. Meisenberg, "A Jensen effect on dysgenic fertility: An analysis involving the National Longitudinal Survey of Youth," *Personality and Individual Differences,* 55 (2013): 279–282.

21 M. Wang, J. Fuerst and J. Ren, "Evidence of dysgenic fertility in China," *Intelligence,* 57 (2016): 15–24.

22 G. Meisenberg and R. Lynn, "On-going trends of human intelligence," *Intelligence,* 96 (2023): 1017098.

23 E. Gardner, M. Neville, K. Samocha et al., "Reduced reproductive success is associated with selective constraint on human genes," *Nature,* 603 (2022): 858–863; M. Kolk and K. Barclay, "Cognitive ability and fertility among Swedish men born 1951–1967: Evidence from military conscription registers," *Proceedings of the Royal Society B: Biological Sciences* (2019):

studies are only on males, so there could be a negative IQ-fertility nexus overall, when females are also analysed. Moreover, it has been argued that Swedish conscription data is unrepresentative. One study found a positive association between income and fertility in the cohorts between 1920 and 1970 but as it was weak this could be driven by positive selection for certain personality traits, which we will explore later.[24] A Swedish cohort study, published in 1988, found that IQ and fertility correlated in Swedish men born up to 1924, after which there was no relationship.[25] Certainly, native Swedish reaction times are consistently slowing,[26] implying declining intelligence. We will explore reaction times in greater depth below.

Evidence for Intelligence Decline and the Flynn Effect

Now let us turn to the evidence that intelligence is in decline. Before we do this, however, we must examine the Flynn Effect. It should also be emphasised that the Flynn Effect—the secular rise in IQ scores across the twentieth century in Western countries—is entirely consistent with intelligence decline, as its putative discoverer,[27] American political scientist James Flynn (1934–2020), himself conceded.[28] As already discussed, intelligence can be conceived of as rather like a pyramid. At the base of the pyramid there are numerous "specialized abilities" that are only very weakly associated with intelligence. The documented rise in IQ

20190359.

24 M. Kolk, "The relationship between life-course accumulated income and childbearing of Swedish men and women born 1940–70," *Population Studies*, (2022).

25 D. Vining, L. Bygren, K. Hattori, S. Nystrom and S. Tamura, "IQ/fertility relationships in Japan and Sweden," *Personality and Individual Differences*, 9 (1988): 931–932.

26 G. Madison, M. A. Woodley of Menie and J. Sänger, "Secular Slowing of Auditory Simple Reaction Time in Sweden (1959–1985)," *Frontiers in Human Neuroscience*, (2018).

27 R. Lynn, "Who discovered the Flynn effect? A review of early studies of the secular increase of intelligence," *Intelligence*, 4 (2013): 765–769.

28 J. R. Flynn, and M. Shayer, "IQ decline and Piaget: Does the rot start at the top?" *Intelligence*, 66 (2018): 112–121.

scores has been found to have been on intelligence abilities at the base of the intelligence pyramid, and, in particular, on the subtest "similarities"—the ability to make associations. This ability is only a very weak measure of intelligence, but due to the fact that it has increased extremely rapidly, and due to the imperfect nature of IQ tests as means of measuring intelligence—certain personality traits, such as impulse control, can also elevate your ability to solve problems—this has shown-up on the IQ tests as an overall IQ rise. Flynn has argued that this occurred because we are an increasingly science-oriented society that increasingly teaches people to think in an analytic way; it makes them don "scientific spectacles."[29]

This process has pushed these intelligence abilities to their phenotypic maximum, overwhelming an underlying decline in actual intelligence. What "actual intelligence" means is difficult to define. Modern conditions of life and especially intense schooling has boosted most cognitive skills, but to very different extents. Generally, abstract analytical skills, like those measured with the Similarities test and the more difficult items on the Raven's Matrices, have been boosted the most, whereas integrative skills (those closer to what we call wisdom) have not risen that much, and may have declined.

Once the phenotypic maximum in similarities is reached, then declining IQ would show up even on the IQ tests, and this is what began to happen in the West with cohorts born around the year 1980, according to some studies. This new trend is being widely referred to as the Negative Flynn Effect, connoting a phenotypic decline in IQ, and there is some evidence that this decline is actually driven by *g*; on the most genetic parts of the IQ test, as we would expect.[30] A study of Norwegian military conscription data averred that the Negative Flynn Effect was entirely environmentally mediated, by comparing siblings.[31] But there are serious

29 J. R. Flynn, *Are We Getting Smarter? Rising IQ in the Twenty-First Century* (Cambridge: Cambridge University Press, 2012), 15.

30 E. Dutton, D. Van der Linden and R. Lynn, "The Negative Flynn Effect: A Systematic Literature Review," *Intelligence*, 59 (2016): 163–169.

31 B. Bratsberg and Ole Rogeberg, "Flynn effect and its reversal are both environmentally caused," *PNAS*, 115 (2018): 6674–6678.

problems with this study, such as the methods employed being extremely opaque.

So, there is no inconsistency between intelligence decline and the Flynn Effect. Indeed, it may be that a hypothetical "Negative Flynn Effect" comes about, leading to accelerated reduction in IQ for environmental reasons. As people become less intelligent for genetic reasons, teachers will be less able to push the IQ of their pupils to its phenotypic maximum, meaning that the pupils' IQ will fall for both genetic and environmental reasons. Other correlates of intelligence are "measurement invariant"—meaning that the validity of the instrument is not interfered with by external changes—and these also evidence an intelligence decline.

Simple Reaction Times

Reaction times are such a reliable proxy for general intelligence that eminent intelligence researchers, such as American psychologist Arthur Jensen (1923–2012),[32] have promoted them as alternatives to pencil-and-paper IQ tests. It is possible to use so-called "simple reaction times" to measure long-term trends in general intelligence because reaction times have been measured since the 1880s and, as we have discussed, they represent an objective correlate of general intelligence. Also, going to school is assumed to raise IQ, due to greater intellectual stimulation, but not to shorten reaction times. Thus, reaction times are measurement invariant across time.

Simple reaction times (sRT) typically involve something like pressing a button as quickly as possible in response to a light being switched on, and measuring the time taken. This procedure usually takes a fraction of a second. While the correlation with IQ is not large, sRTs have the huge advantage of being objective and quantifiable physiological measures. It might be asked how we can measure general intelligence in Victorian England. The answer is that Galton measured simple reaction time, which is a measure of endophenotypes (such as nerve conduction velocity) that are considered by some prominent researchers to be funda-

32 A. R. Jensen, *Clocking the Mind: Mental Chronometry and Individual Differences* (New York: Elsevier, 2006).

mental determinants of *g*. A critical strength of these measures is that their *meaning* does not change over time. The low *g*-loading of measures like simple reaction time is therefore not relevant to their ability to reliably track the change in the underlying *g* over time. The critical point is that they exhibit the property of *measurement invariance*.

A survey of historical reaction time data demonstrates something striking.[33] There was a significant slowing of sRTs from the 1880s until the late twentieth century. It should be stressed that the instruments used to measure reaction times in the 1880s are accepted as having been perfectly adequate for the job.[34] This was then replicated and confirmed.[35] This replication study furthermore found, using sRTs, that the decline in *g* had been around 1 IQ point per decade between 1885 and the year 2004. That is about 10 points in a century, and probably more over the past two hundred years. English medical doctor Bruce Charlton and I have explained that to put this in perspective, 15 points would be approximately the difference in average IQ between a low level security guard (85) and a police constable (100), or between a high school science teacher (115) and a biology professor at an elite university or an engineer (130).[36] In other words, in terms of intelligence, the average Englishman from about 1850 would be roughly at the 85th percentile of the population in the year 2000—and the difference would be even larger if we extrapolated back further towards about 1800 when the Industrial Revolution began to initiate massive demographic changes in the British

33 I. W. Silverman, "Simple reaction time: it is not what it used to be," *American Journal of Psychology,* 123 (2010): 39–50.

34 M. A. Woodley, J. te Nijenhuis and R. Murphy, "Were the Victorians cleverer than us? The decline in general intelligence estimated from a meta-analysis of the slowing of simple reaction time," *Intelligence,* 41 (2013): 843–850.

35 M. A. Woodley of Menie, J. te Nijenhuis and R. Murphy, "The Victorians were still faster than us. Commentary: Factors influencing the latency of simple reaction times," *Frontiers in Human Neuroscience,* 9 (2015): 452.

36 E. Dutton and B. G. Charlton, *The Genius Famine* (Buckingham: University of Buckingham Press, 2015), 158–159.

population. New studies are replicating this finding. A study in Sweden, for example, based on a sample of more than 7,000 people, found that simple audio reaction times had slowed by between 3 and 16 milliseconds between 1959 and 1985.[37]

In everyday terms, the meaning of these numbers is stark. They mean that we live in an increasingly over-promoted society, as those at the top fail to reproduce and those at the bottom substantially reproduce. Bruce Charlton and I set out the full impact of these numbers in *The Genius Famine:*

> [...] the academics of the year 2000 were the school teachers of 1900, the school teachers of the year 2000 would have been the factory workers (the average people) of 1900, the office workers and policemen of the year 2000 were the farm labourers of 1900, those who were around 10 to 15 IQ points below average at that time. The low-level security guards and shop assistants of the year 2000 were probably in the workhouse, on the streets, or dead in 1900. The substantial long-term unemployed or unemployable, the dependent "underclass" of the year 2000, simply didn't exist in 1900. And even this estimate is ignoring the expansion of education since 1900, which expanded the middle class occupations and would, in itself, reduce the average intelligence of academics and teachers and even shop assistants in 2000 compared to what they would have been in 1900.[38]

Colour Discrimination

Colour discrimination—the ability to distinguish between ever more subtle shades of colour—is declining as well. Sensory discrimination correlates with *g* because the more acute your sensory discrimination is, the better able you are to notice subtle differences between physical quantities which can help you to

37 G. Madison, M. A. Woodley of Menie and S. Sänger, "Secular slowing of auditory simple reaction time in Sweden, (1959–1985)," *Frontiers in Human Neuroscience,* 10 (2016): 407.

38 E. Dutton and B. Charlton, *The Genius Famine, op. cit.,* 158–159.

solve problems more efficiently. This is why studies have found the correlation between general intelligence and general discriminative ability to be as high as 0.92 in some cases.[39] It follows that the more intelligent people are, the better they will be able to discriminate among increasingly subtle differences in colour.

A 2015 study explored four standardization studies conducted between the 1980s and 2000 that drew upon the Farnsworth-Munsell 100-hue Colour Perception Test, which was developed in 1943. This test was found to correlate with IQ in the 1960s. The participants have to physically arrange a series of 85 caps, each of a very subtly different hue, along a spectrum defined by two clear end caps such as blue and green or pink and purple. Participants are awarded an error penalty for each cap that is in the wrong order on the spectrum, allowing a quantification of their colour discrimination abilities. It was found that, across the four studies, colour discrimination ability had significantly declined, at the equivalent rate of 3.15 IQ points per decade—even after controlling for the IQ of the countries from which the participants were drawn (Belgium, Finland, the UK and the US) and also participant age. Clearly, this is what we would expect if our hypothesis were correct.[40]

Use of High-Difficulty Words

As discussed earlier with regard to the Surfside Building Collapse, the more intelligent people are, the larger are their vocabularies, and the more likely they are to use difficult words. Intelligent people are more prone to using "high order", or at least unusual words. This is congruous with the association between *g* and colour discrimination. The more intelligent people are, the more able they are to perceive subtle differences; differences which re-

39 I. J. Deary, J. P. Bell, M. J. Campbell and N. D. Fazal, "Sensory discrimination and intelligence: Testing Spearman's other hypothesis," *American Journal of Psychology*, 117 (2004): 1–18.

40 See, M. A. Woodley of Menie and H. B. F. Fernandes, "Showing their true colours: Secular declines and a Jensen effect on colour acuity—more evidence for the weaker variant of Spearman's other hypothesis," *Personality and Individual Differences*, 88 (2015): 280–284.

quire slightly different words in order to accurately encapsulate them. Part of the linguistic intelligence dimension of the IQ test is vocabulary and the ability to understand subtle differences in the meaning of words. Scores on vocabulary tests are *very* highly *g*-loaded and highly heritable.[41]

To test what was happening with vocabulary, a study in 2015 examined historical changes in the frequency with which words from the highly *g*-loaded WORDSUM test were employed across 5.9 million texts published between 1850 and 2005.[42] It also examined the association between WORDSUM scores and completed fertility; how many children you have had by middle age when, typically, you don't have any more. They found that knowledge of words with higher difficulties (those that are harder to learn and use correctly) had stronger negative correlations with completed fertility, and their use declined over time. By contrast, less difficult words and less strongly selected words increased in use over time—an effect that was predicted by rising literacy.

More recently, another study has taken this further in a way that clearly illustrates the accuracy of our model. Google's Ngram Viewer (a massive text archive of scanned books, newspapers, scientific journals and other printed materials) includes texts that go all the way back to the sixteenth century. So, drawing upon this, they analysed changes in the use of the four very high difficulty WORDSUM words over time between the sixteenth century and the modern day. They found that the use of these words *increases* from the sixteenth century up until the early nineteenth century and then goes into decline. This is precisely what our model of the rise and fall of Western intelligence would predict, if indeed the usage patterns of these words among those

41 K. J. Kan, J. Wicherts, C. Dolan and H. Van der Maas, "On the nature and nurture of intelligence and specific cognitive abilities: the more heritable, the more culture dependent" *Psychological Science*, 24 (2013): 2420–2428.

42 M. A. Woodley of Menie, H. Fernandes, A. J. Figueredo and G. Meisenberg, "By their words ye shall know them: Evidence of genetic selection against general intelligence and concurrent environmental enrichment in vocabulary usage since the mid-19th century," *Frontiers in Psychology*, 6 (2015): 361.

who contribute to literature really does reflect their underlying level of *g*.[43]

This system also allows us to estimate roughly how intelligent we are compared to people in the past. Based on the usage frequencies of these WORDSUM words, we currently have about the same level of *g* as people in the early eighteenth century—a couple of generations before the Industrial Revolution. It must be remembered that WORDSUM is likely to be subject to the effects of more widespread education (recall that the usage of easy words is actually increasing), which may make our vocabulary level artificially high in relation to our underlying general intelligence. So, putting these influences aside, it is likely that we have regressed considerably further, genetically, than the early eighteenth century.

Backward Digit Span

Another proxy for *g* is working memory, or the capacity to manipulate information committed to memory for the purpose of solving problems. More intelligent people tend to be better at this. This makes sense because if you have a good working memory, the amount of information that you can handle will be greater, allowing for more complex problems to be solved. This ability is reflected in measures such as "Digit Span," where the subject is presented with a list of digits and must immediately repeat them back from memory. If they can do this successfully, then they move onto a longer list. The length of a list that the subject can recall successfully is their "Digit Span." They can be asked to recall the digits in the order in which they were given ("forward"), which gives a measure of short-term memory, or they can be asked to recall them in the opposite order. The latter is known as their "Backward Digit Span" and is a measure of working memory. Clearly, remembering numbers in reverse order is likely to be more cognitively demanding and is therefore a much better measure of *g*.

43 M. A. Woodley of Menie, A. J. Figueredo, M. A. Sarraf et al., *The Rhythm of the West: A Biohistory of the Modern Era, AD 1600 to the Present* (Washington, DC: Scott Townsend Press, 2017).

In a reanalysis of previously published data, covering the period 1923–2008, it was found that forward digit span (short-term memory) had slightly improved over this period. However, backward digit span (working memory) had declined—equating to an IQ loss of 0.16 points per decade. In other words, we have become better at the less *g*-loaded memory task and worse at the more *g*-loaded task over a period of 85 years.[44]

Levels of Genius

There are many other proxies for intelligence which are also declining, including spatial perception[45] and creativity.[46] But, most importantly, levels of genius, of major scientific innovation per capita, are in decline. As discussed already, per capita rates of genius and the macro-innovations for which they are responsible have been declining since the early to middle nineteenth century.[47] This, of course, is the height of the Industrial Revolution. Then they go into decline. This can be seen in Figure 2.

44 M. A. Woodley of Menie and H. Fernandes, "Do opposing secular trends on backwards and forwards digit span evidence the co-occurrence model? A comment on Gignac (2015)," *Intelligence,* 50 (2015): 125–130.

45 J. Pietschnig and G. Gittler, "A reversal of the Flynn effect for spatial perception in German-speaking countries: Evidence from a cross-temporal IRT-based meta-analysis (1977–2014)," *Intelligence,* 53 (2015): 145–153.

46 K. H. Kim, "The creativity crisis: The decrease in creative thinking scores on the Torrance Tests of Creative Thinking" *Creativity Research Journal,* 23 (2011): 285–295.

47 C. Murray, *Human Accomplishment: The Pursuit of Excellence in the Arts and Sciences, 800 BC to 1950* (New York: Free Press, 2006).

Figure 2: Per Capita Major Innovations, 1440–2000

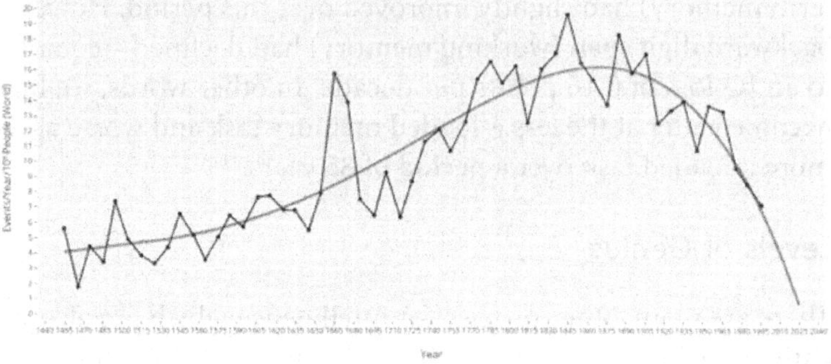

It might be argued that this decline in "recognized important innovations" has occurred because we have already picked all of the "low-hanging fruit." But this misunderstands how problem-solving works. Problem-solving involves making connections between pieces of already available information, so that, when it comes to solving problems, there is no such thing as "low-hanging fruit." When inventions, which we are now used to, were first innovated, they were regarded as extraordinary breakthroughs; geniuses having made connections which nobody had yet been sufficiently intelligent to make. It might also be averred that though intelligence is declining, the population is rising, meaning any intelligence decline is more than offset by a rise in the absolute numbers of highly intelligent people. However, it has been shown that our intelligence decline is so rapid that this is simply not the case, not least because a decline of just 3 IQ points almost halves the percentage of people with an IQ over 130.[48] Moreover, there is substantial evidence that the average IQ of a society, as well as percentage of it that is part of the "smart fraction" (who have an IQ of over 130) very strongly influences its ability to maintain all aspects of civilization.[49]

48 M. A. Woodley and A. J. Figueredo, *Historical Variability in Heritable General Intelligence: Its Evolutionary Origins and Socio-Cultural Consequences* (Buckingham: University of Buckingham Press, 2013).

49 See H. Rindermann, *Cognitive Capitalism: Human Capital and the Well Being of Nations* (Cambridge: Cambridge University Press, 2018).

A Decline in the Frequencies of Variants in the Genome Associated with *g*

So far, we have looked at phenotypic evidence, which indicates (quite strongly) that *g* is declining and doing so for largely genetic reasons. A hugely important study, led by a Chinese researcher called Augustine Kong was published in 2017.[50] In this study, this team identified a large number of genetic variants, which collectively predicted both educational attainment and *g*. They called this set of variants POLYEDU (polygenic score for educational attainment). This team investigated the effect of this polygenic score on the reproductive histories of 109,120 Icelanders, and the impact of this history on the Icelandic gene pool over time. They demonstrated that those who had higher POLYEDU had delayed reproduction and fewer children than did Icelanders carrying lower POLYEDU. So far, this result is somewhat consistent with those of the previously discussed studies that used polygenic scores for educational attainment to predict fertility outcomes. However, Kong and his team went one step further. Based on a sample of 129,808 Icelanders born between 1910 and 1990, they found that the average POLYEDU—the average percentage of the population with genes that predict high educational attainment—had been declining at a rate of roughly 0.010 standard units per decade, which, they noted "is substantial on an evolutionary timescale." They added that "because POLYEDU only captures a fraction of the overall underlying genetic component, the latter could be declining at a rate that is two to three times faster."

This observed decline over decades in the population's levels of POLYEDU was found to be highly consistent with the decline predicted using the negative association between POLYEDU and fertility, and the positive association between POLYEDU and age at first birth (recall that those with high IQ don't simply produce fewer children, they produce them later in life). Kong and his colleagues even went so far as to estimate the IQ-loss that should

50 A. Kong, M. Frigge, G. Thorleifsson et al., "Selection against variants in the genome associated with educational attainment," *Proceedings of the National Academy of Sciences, USA*, 114 (2017): E727-E732.

result from this process—0.3 points per decade. It is important to note that in arriving at this estimate they employed a *very* low value of the heritability of IQ (0.3) in their formula. As was discussed earlier, the actual heritability of IQ (as determined using twin studies, which have the advantage of capturing *all* genetic variants that go into a given trait) is likely to be substantially higher (closer to 0.8). When Kong and his colleagues' numbers are adjusted to take this into account, the decline increases to around 0.8 points per decade—meaning that the population of Iceland may be losing IQ at a rate very close to a whole point per decade.[51]

We have already seen that education level is strongly genetically correlated with *g*. So, *g* should have declined between 1910 and 1990 among the people of Iceland. The average Icelander born in 1910 was cleverer than the average Icelander born in 1990 in terms of heritable general intelligence. This decline in Iceland has furthermore occurred exclusively for genetic reasons and the key reason is the low and postponed fertility of the highly intelligent in Iceland, especially the highly intelligent women.

What About Pollution?

Some researchers have argued that declining general intelligence is caused by environmental pollution. In her 2014 book *Losing Our Minds,* British biologist Barbara Demeneix argues that the rise in pollution from certain kinds of plastic and other modern pollutants, coupled with the effects of "ever present neurotoxins" in the environment and in our diets (such as lead, mercury and alcohol), significantly explains the evidence for IQ (specifically *g*) decline.[52] In particular, these neurotoxins act as endocrine disruptors, leading to epigenetic changes—alteration of genetic expression—in the early stages of foetal development that have resulted in reduced *g*.

51 Woodley of Menie, Figueredo, Sarraf et al., *The Rhythm of the West, op. cit.*

52 B. Demeneix, *Losing Our Minds: How Environmental Pollution Impairs Human Intelligence and Mental Health* (Oxford: Oxford University Press, 2014).

A team of researchers have tested Demeneix's neurotoxin theory. They created a model in which they predicted the decline in general intelligence using a neurotoxin factor (comprised of environmental measures of lead, mercury and persistent organic pollutants, along with per capita alcohol consumption for the US and UK) and a polygenic score factor. This was comprised of genetic variants which predict *g* sampled from individuals of different ages based on birth year—sourced from the aforementioned Iceland study and also from a smaller study presenting similar findings for cohorts born between 1919 and 1955 in the US. They found that changes in polygenic scores predicted around 25 per cent of the variance in the change in *g* over time. Changes in neurotoxic exposure did not significantly predict any of the variance in the change, however.[53] The neurotoxin theory simply does not seem to be supported by the data.

The Decline of National IQs and Immigration

So, it is clear that intelligence is in decline, and this is primarily for genetic reasons. Based on per capita levels of genius, Western intelligence peaked around 1870, and we have now returned to the levels of per capita genius of which we were capable in about the year 1600. Based on the use of high difficulty words, we have returned to where we were around 1650.

"National IQs" have been heavily criticised and, thus, recalculated to take these criticisms into account. They correlate with the originals at close to 0.9.[54] National average IQ is associated with all aspects of civilization and often strongly so. National IQ is associated with educational level, literacy, low corruption, law and order, political stability, life expectancy, health and much else.[55] Moreover, there has been shown to be a correlation of 0.9

53 M. A. Woodley of Menie, M. Sarraf, M. Peñaherrera-Aguirre et al., "What caused over a century of decline in general intelligence? Testing predictions from the genetic selection and neurotoxin hypotheses," *Evolutionary Psychological Science*, 4 (2018): 272–284.

54 R. Lynn and D. Becker, *The Intelligence of Nations* (London: Ulster Institute for Social Research, 2019).

55 R. Lynn and T. Vanhanen, *Intelligence: A Unifying Construct for the Social Sciences* (London: Ulster Institute for Social Research, 2012).

between the prevalence of alleles associated with very high intelligence and national IQ.[56] So a decline in national average IQ would eventually lead to the decline of civilization, and we know that national intelligence is declining.

National IQs are declining in many Western countries, among the native population, as noted above. The consequences of this are likely to be worsened by the fact that most immigrant groups in Western countries have lower IQs—Indians in the UK have an IQ of 96, with Bangladeshis and Pakistanis it is 86, and it is 102 with Chinese[57]—and higher fertility rates than do natives.[58] For example, the average IQ of immigrants in Denmark is 89, compared to 100 for Danes.[59] This will potentially reduce the IQ even of natives as society becomes less able to push native IQs to their phenotypic maximum due to factors such as declining educational standards, with schools needing to accommodate decreasingly intelligent pupils.

Eating Into the Capital

It might be argued that the intelligence decline that we have examined is not consistent with everyday life, in which society becomes ever more technological, and living standards seem to improve across generations. It can be countered that the innovations of the Industrial Revolution were so enormous and boosted living standards so substantially, that the Industrial Revolution was akin to a huge amount of capital.[60] With the Industrial Revolution, we managed to make an enormous amount of money. We became quite spectacularly rich, because of the brilliant inventions we came up with. This permitted us an extremely

56 D. Piffer, "Correlation between PGS and environmental variables," *RPubs* (2018), https://rpubs.com/Daxide/377423.

57 R. Lynn and H. Cheng, "Recent data for majority and racial minority differences in intelligence of 5 year olds in the United Kingdom," *Intelligence*, 41 (2013): 452–455.

58 Lynn, *Dysgenics, op. cit.*, 272.

59 E. Kirkegaard, "Predicting Immigrant IQ from their Countries of Origin, and Lynn's National IQs: A Case Study from Denmark," *Mankind Quarterly*, 54 (2013): 151–167.

60 Dutton and Charlton, *The Genius Famine, op. cit.*

high standard of living: the stereotypical American Dream of the mansion, the country retreat, the jet-setter lifestyle and so on. But, thereafter, our rate of inventiveness slowly decreased. At first this didn't matter, because of the huge interest that our capital was generating; the capital was making more capital, with very little input on our part. Our input was just minor tinkering, little inventions here and there. But, eventually, our rate of inventiveness completely dried up, and once this happened, we were no longer making any new money. So, we had to start living off the capital and, in order to sustain this, standards of living would necessarily decline.

We can think of a pyramid of invention. The inventor of the car is cleverer than the developer of the car. He is smarter than the person who can manufacture, and fix the car, while he is more intelligent than the person who can maintain the car, and he in turn, is brighter than the man who can merely operate the car. Each generation, the top level is boiling off, meaning we are living in an increasingly over-promoted society. Eventually, we will decline to a level where we can no longer manufacture cars. Then we will run out of cars, and, on that measure, civilization will go into reverse.

Of course, some areas, knowledge and skills are cumulative. Science and engineering are of this type, and it is here where high levels of performance can be sustained for a long time in spite of declining intelligence. In other areas, intelligence is expressed more directly in daily practice rather than being formalized in the form of teachable facts and procedures. Politics, management, and education belong here, and especially the creation of the value systems that form the foundations of every human society. Here, performance depends directly on the intelligence of individual politicians, managers, teachers, and intellectuals, not on regurgitated and applied book knowledge based on past innovations. Therefore, the pattern of civilizational decline that we might predict is one where technology and industrialism can be preserved more-or-less intact through several centuries, although very complex but non-essential activities like space exploration will be abandoned. However, these declining human societies will be vulnerable to maladaptive ideologies and political failures with rampant exploitation, wars, and civil unrest.

During this period, economic decline will be driven not by the loss of technologies, but by political and management failures that prevent their efficient use.

However, as we will discuss, there are likely to be refuges of civilization during this decline, because of the inevitability of intelligent people migrating away to richer places, where other intelligent people exist. Intelligence has consistently been shown to predict migration. This is because intelligent people are more open-minded to new possibilities (which allow them to better solve problems), they are more future-oriented and better prepared to make sacrifices now for a better future and they are wealthier and more organised, meaning that they are more able to migrate. This is the "islands of prosperity" future, where a few highly attractive and prosperous countries maintain themselves by sucking the talent out of the rest of the world.[61]

We have not reached this situation yet, but it can, perhaps, be seen in a decreasing ability to solve novel and difficult problems, of the kind that would have been relatively easy for the geniuses of 1850, when our cognitive elite were at their IQ peak. In this regard, it has been shown that the number of scientists required to double the speed of technology has itself doubled since 1979.[62] But it is not merely intelligence that is in decline: so is "moral character", and this is the issue to which we will now turn.

61 For a more detailed discussion of this see Edward Dutton and J. O. A. Rayner-Hilles, *The Past is a Future Country: The Coming Conservative Demographic Revolution* (Exeter: Imprint Academic, 2022).

62 N. Bloom, C. I. Jones, J. Van Reenen and M. Webb, "Are Ideas Getting Harder to Find?" *American Economic Review*, 110 (2020): 1104–1144.

Five

The Rise of Pro-Social Personality, Mental Health, and Religiosity

Risking Your Life for Your Country

In summer 1914, the anthropologist-to-be J. D. Unwin (1895–1936) had just been accepted into Oxford University with a scholarship to read Classics and was about to start his degree when the Great War broke out. Conscription wasn't introduced until 1916, but, led by a burning desire to fight for his country, Unwin immediately joined up. He was wounded three times: at the opening Battle of the Somme in 1916, at a later Battle of the Somme in 1917 and, finally, in 1918 in Normandy. This third set of wounds was so severe that Unwin was hospitalized for two months and nearly died. Indeed, the wounds directly contributed to his premature death almost two decades later.[1]

Only a week prior to receiving these wounds, Unwin had come very close to laying down his life for his country. The captain

> was second in command of a company of tanks operating in front of the infantry. Realising that there was a danger of his tanks losing direction owing to the thick mist, Capt. Unwin went ahead on foot in front of the tanks and infantry, under heavy machine-gun and

1 For a biography of J. D. Unwin, see, E. Dutton, "Introduction: A Critique of *Sex and Culture,* and the Life of J. D. Unwin," in J. D. Unwin, *Sex and Culture* (Perth: Imperium Press, 2023).

> rifle fire, to locate the crossings of the railway which were his objective. He then returned to his tanks, and again walking in the open just ahead of them, guided them successfully to their objectives. Capt. Unwin showed a total disregard for his personal safety, and it was undoubtedly due to his gallantry that the tanks reached their objectives up to time.

For this "conspicuous gallantry," Unwin was awarded the Military Cross.[2]

Unwin was far from alone in engaging in this kind of self-sacrificial behaviour. George Pitt-Rivers (1890–1966), later an anthropologist and eugenicist, also joined up right at the beginning of the war. In November 1914, he was badly wounded by German machine gun fire in the left leg and had to use a walking stick for the rest of his life.[3] It seems inconceivable that young men would have rushed to fight for Britain if such a war had broken out in summer 2014, as by then the native English were increasingly a different kind of people.

Selection for Pro-Social Personality and for Mental Health

We have noted thus far that, until the Industrial Revolution, we were under harsh Darwinian selection for intelligence. However, we were also under intense selection for a certain kind of modal personality. Personality is our general way of being. Differences in personality predict differences in how people will respond in a certain situation. As Bruce Charlton and I put it in *The Genius Famine*:

> How close to you does a car moving at a certain speed have to be before you judge that it is too dangerous to cross the road in front of it? How many annoying things have to happen to you in a day before you lose your temper and raise your voice? How strongly do

2 *Tank Corps Honours and Awards, 1916–19* (Midland Medals, 1982).

3 B. W. Hart, *George Pitt-Rivers and the Nazis* (London: Bloomsbury, 2015), 24.

the emotions of others impact how you feel?[4]

Different people will give different answers to these questions, and they will do so because they have different kinds of personality. Many different models of personality exist, and psychologists attempt to measure personality by asking people to describe themselves on a scale, often in relation to certain statements such as, "I am a caring person." These different narrow aspects of personality can be reduced to five essential characteristics that are substantially independent of each other. These are known as the "Big Five"[5] and they are widely accepted in psychology as being the best way to conceive of personality differences. They can be remembered with the acronym "OCEAN." They are:

(1) Extraversion. Extraversion is feeling positive feelings strongly. It is associated with being outgoing and enjoying the company of others. "Introversion" is experiencing these feelings weakly. It is associated with being reserved and not needing the company of others.

(2) Neuroticism. Neuroticism is feeling negative emotions strongly. A person high in Neuroticism is emotionally unstable and prone to mood-swings, anxiety, and depression. A person who was low in Neuroticism would be mentally stable and would not let negative events "get them down." They would tend to "always look on the bright side of life," especially if they were also high in Extraversion.

(3) Conscientiousness. Conscientiousness refers to impulse control. People who are high in this are self-disciplined, hard-working and obey the rules. People who are low in impulse control are lazy, spontaneous, emotional, and do not as stringently abide by the rules.

(4) Agreeableness. Agreeableness refers to those who are

4 E. Dutton and B. G. Charlton, *The Genius Famine* (Buckingham: University of Buckingham Press, 2015).

5 See: P. T. Costa and R. R. McCrae, *Revised NEO Personality Inventory (NEO-PI-R) and NEO Five-Factor Inventory (NEO-FFI) Manual* (Odessa, FL: Psychological Assessment Resources, 1992).

high in empathy and who are pro-social. They deeply care about, and empathize with, how others feel. They are altruistic. Those who are low in Agreeableness are selfish and even callous.

(5) Openness-Intellect: Openness-Intellect is a broad domain of personality, which involves intellectual curiosity and a preference for novelty, political liberalism, creativity, aesthetics, and unusual psychological experiences. The opposite pole of this domain is characterized by lack of interest in novelty, a preference for conservative politics and for pragmatism. Openness-Intellect weakly, but significantly, correlates with intelligence, at 0.3. This is because it, or the "Intellect" domain, is measuring some of the same things.[6]

These five personality traits are (except for Openness-Intellect) regarded as substantially independent of IQ scores and they predict important life outcomes. For example, high Conscientiousness as a child predicts greater success in education and employment. In other words, it is associated with attaining high socioeconomic status.[7] High Neuroticism predicts problems with anxiety and depression. It is correlated with suffering from depression at 0.8.[8] High extraversion, interestingly, predicts premature death because extraverts are more likely to take risks in pursuit of positive feelings. They do so because they will enjoy the resulting payoff so much more. One study found that high extraversion leads to a three-fold increase in the risk of early death.[9] Extraversion predicts poor health, because of the extra

6 See, D. Nettle, *Personality: What Makes You Who You Are* (Oxford, Oxford University Press, 2007).

7 M. Almlund, A. Duckworth, J. Heckman, and T. Kautz, "Personality, psychology and economics" in S. Hanushek, S. Machin and L. Woesmann, (Eds.). *Handbook of the Economics of Education* (Amsterdam: Elsevier, 2011).

8 Nettle, *Personality, op. cit.*

9 H. S. Friedman, J. Tucker, C. Tomlinson-Keasey et al., "Does childhood personality predict longevity?" *Journal of Personality and Social Psychology*, 65 (1993): 176–185.

enjoyment that will be found in, for example, sweet or fatty food. High Agreeableness predicts having lots of friends. Low Agreeableness is associated with criminality.[10]

The General Factor of Personality

British-Canadian psychologist J. Philippe Rushton (1943–2012) is infamous for his involvement in a series of controversies over race differences in intelligence and personality, one of which, in 1989, led to his nearly being prosecuted under Canada's stringent anti-free speech laws.[11] However, what is less well known is that Rushton, among others, showed that the Big Five (and Big Three) are all co-correlated, and could all, therefore, be reduced down to a single personality variable that he called the General Factor of Personality (GFP). GFP can be conceptualized as *social effectiveness*, the ability to effectively read people and social situations; to a certain extent it may even be regarded as similar to the "emotional intelligence" discussed earlier. GFP underlies the more specific personality traits, just as general intelligence or "*g*" underlies all the different cognitive abilities. With personality, you have the "aspects," numerous very narrow traits, such as "optimism" or "fearfulness." These can be reduced down to the Big Five and these, in turn, yield a Big Two. The Big Two are *Stability* (Conscientiousness, Agreeableness and Emotional Stability) and *Plasticity* (Extraversion and Openness-Intellect). The Big Two correlate. So, we end up with the *g* factor of personality—the "General Factor of Personality."[12]

A person with high GFP would be extraverted, empathetic, concerned with the feelings of others, conscientious and self-disciplined in pursuit of socially acceptable goals, have stable emotions, and he would be open to new ideas. The existence of the GFP is also why people say things like "she has a nice personality,"

10 See: Nettle, *Personality, op. cit.*

11 See E. Dutton, *J. Philippe Rushton: A Life History Perspective* (Oulu: Thomas Edward Press, 2018).

12 J. P. Rushton and P. Irwing, "A General Factor of Personality from two meta-analyses of the Big Five," *Personality and Individual Differences,* 45 (2008): 679–683.

or "he's not a very nice person." It can be clearly seen in fairy tales and children's television programmes. Characters are, in general, either "goodies" or "baddies." They are either "those who give" or they are "those who take." There is, perhaps, some greater degree of complexity: with some "goodies" being less fearful than others and some "baddies" being cleverer than others. But this essential divide, really between "good" and "evil," seems to be something we very easily intuit. People seem to intuitively understand that there is one core personality trait with socially desirable and undesirable dimensions. Paying attention to this in choosing mates and allies likely would have had significant evolutionary advantages. The GFP predicts likeability and employability, and is strongly correlated with Emotional Intelligence.[13] This is the ability to understand yourself and others.

How Personality Develops and Why there are Differences

Personality develops throughout the lifespan, just as intelligence does.[14] Anyone who has ever had a four-year-old child will notice that, in comparison to a child of 10, the four-year-old will have lower impulse control and is more likely to have a tantrum, is less considerate to his or her friends, and is far more likely to become scared of something or extremely upset. We equally see that the elderly tend towards being non-aggressive but highly resistant to change. They become "set in their ways" and don't want to experiment with anything new. The internet dominates most of our lives these days, and most of us simply couldn't imagine life without it. Personally, I find it mind-boggling to realize that I didn't send an email until I was an undergraduate in 1999 and I didn't start using the internet for anything other than checking emails or reading the news until late 2005. But many very elderly

13 C. Just, "A review of literature on the general factor of personality," *Personality and Individual Differences,* 50 (2011): 765–771.

14 See: C. Soto, O. John, S. Gosling, and J. Potter, "Age differences in personality traits from 10 to 65: Big Five domains and facets in a large cross-sectional sample," *Journal of Personality and Social Psychology,* 100 (2011): 330–348.

people in the UK at the time of writing have never used a computer at all, let alone the internet. These things didn't exist when they were middle aged, and so there is a tendency not to want to have anything to do with them. Similarly, many people born in the 1880s but still alive in the 1960s could cope with a radio but never purchased a television or, for that matter, a telephone.

Conscientiousness increases with age up until puberty, where people regress a bit (see below), and then from early adulthood it continues increasing. Exactly the same is true of Agreeableness; people become more Agreeable as they get older. This may well be why children tend to prefer people of their grandparents' generation to people of their parent's generation. Older people are objectively more altruistic. In males, Neuroticism decreases with age, while in females it decreases with age until adolescence. At that point, it increases. This can be seen most obviously among students, where many females in particular will tend to suffer from mental health problems of various kinds that they will often have recovered from by the time they are in their mid-twenties. Neuroticism, in women, begins to decline in early adulthood. For this reason, there is considerable personality variation on this measure between the sexes. Women score higher in Neuroticism then men and are more prone to anxiety.[15]

It is fascinating that the move towards higher Conscientiousness as one gets older experiences a blip during adolescence. Perhaps there is some evolutionary benefit to this. It may be that young people become less inhibited and therefore potentially more creative. This would mean that they'd be more likely to attract the attention of a mate or produce some useful innovation for their group. Such behaviour might also make them more likely to break or loosen the bond with their parents—better able to set out on their own life path. Once people reach adulthood their personalities, compared to others of their own age, are quite stable. One study found that the correlation between personality scores over six years on the same adult sample was 0.85.[16] It is

15 Soto, John, Gosling, and Potter, "Age differences in personality traits from 10 to 65," *op. cit.*

16 P. Costa and D. Arenberg, "Enduring dispositions in adult males," *Journal of Personality and Social Psychology*, 38 (1980): 793–800.

equally noteworthy that Neuroticism increases among females, though not males, during adolescence. A possible explanation might be that it is females who, in prehistory at least, would have borne more of the consequences of any sexual encounter, insomuch as there would be the potential for females to become pregnant. Higher Neuroticism, leading to a greater inclination to worry, would be useful at that age, because pregnancy would be one of the things a female would worry about, leading her to be very careful regarding which men she became sexually involved with.

There are clear personality differences between males and females. These are consistent with stereotypes about women entering caring professions and men being more aggressive and inclined to fight and, specifically, to fight for status. By the time they reach adulthood, women are higher in Agreeableness and Conscientiousness than men. This would make women better carers for children and men higher in drive and aggression. Women are also higher in Neuroticism than men, which would mean a greater proneness to suffering from stress and worry in any given situation, possibly aiding them in successfully raising children. They are higher in Extraversion than men are. Women are more likely to be outgoing and have an active social life. The differences on Openness-Intellect are at the aspect level. Women are higher in aestheticism; men are higher in "Intellect."[17] So, you would predict that women would be more interested in the Arts and men more interested in the sciences.[18]

Personality traits, including the GFP, have been shown to be in the region of at least 0.5 heritable. As with "intelligence," this is based mainly on twin and adoption studies. Since the heritability of personality is less than one, some combination of chance and the environment clearly must impact the kind of personality you have, but within certain genetic limits. An unstable, dangerous

17 Y. Weisberg, C. DeYoung and J. Hirsch, "Gender differences in Personality across the ten aspects of the Big Five," *Frontiers in Psychology,* 2 (2011): 178.

18 Indeed, this is precisely what Baron-Cohen's model of the "extreme male brain" has found. See: S. Baron-Cohen, "The extreme male brain theory of autism," *Trends in Cognitive Sciences*, 6 (2002): 248–254.

childhood will increase mental instability. Those who experience it will learn to see the world as a perilous place, and this may have a lasting effect on their behaviour. For instance, when childhood is unpredictable and dangerous, children will tend to "live for the now," displaying lower Conscientiousness, and they may be suspicious of other people, leading to lower Agreeableness.[19] Girls who have grown up in sexually unstable households, with no father present, seem to adopt a short-term sexual strategy, having children with a variety of men who are chosen for being tough rather than their ability to remain committed to a relationship and provide resources or care for the children.[20]

Life History Strategy and Personality

Clearly, we would have been under harsh selection for a certain kind of personality, and this would have been more intense in some environments than others, specifically environments that promote what is called a "slow Life History Strategy." Life History Strategy (LHS) refers to a spectrum with fast LHS (*r*-strategy) at one end and slow LHS (*K*-strategy) at the other. If the ecology is highly unstable, meaning you could be killed at any moment, then people tend to adopt a fast LHS. In essence, this means that they "live fast, die young." They invest most of their energy in having sex, they have a large number of sexual partners, and they tend to have a large number of children, which they invest very little energy in, as some are bound to survive. In this ecology, social bonds are weak, because its unpredictability means that they may never be reciprocated. People have low impulse control, because basic needs are met, meaning they need not plan much, and they must be able to act lethally at a moment's notice. Equally, there is little selection for mental stability, as pathological violence, motivated by intense negative feelings, may actually

19 D. Simonton, "Varieties of (scientific) creativity: A hierarchical model of domain-specific disposition, development, and achievement," *Perspectives on Psychological Science*, 4 (2009): 441–452.

20 D. Bugental, R. Corpuz and D. Beaulieu, "An Evolutionary Approach to Socialization," in J. Grusec and P. Hastings, (Eds.), *Handbook of Socialization: Theory and Research* (New York: Guildford Publications, 2014).

mean that you are more likely to survive. The kind of people that this ecology produces can be seen in the Yanomamö of Venezuela, whom we met earlier.[21]

As the ecology becomes more stable, it reaches the maximum carrying capacity for the species, and so members of the species begin to compete with each other for survival. In that the ecology is now predictable and also competitive, if you adopt a fast LHS and invest nothing in any of your children, it is likely that they will all simply die young, for example due to insufficient protection during winter. As such, the society moves towards a slower LHS, in which resources are directed away from copulation and towards "nurturing" a smaller number of offspring, so that they are able to survive the predictable yet harsh ecology. They are more likely to be able to do this if they are higher in Agreeableness, Conscientiousness, and intelligence, so these qualities are part of a slower Life History Strategy. And if they are higher in these qualities then their group will have strong internal bonds and they will be more positively ethnocentric,[22] meaning their group is more cohesive and more likely to survive in the battle of group selection. In such an environment, life itself slows down. Puberty, for example, is reached later, with a longer childhood allowing the offspring to better learn the more complex information necessary in order to survive as an adult in a harsher and more competitive environment.

Religiousness, it should be observed, has been specifically highlighted as an example of a relatively slow LHS. It is a marker of strong in-group bonds, high General Factor of Personality, strong investment in the nurture of offspring, the following of group rules and thus cooperativeness and altruism and a strong interest in socialising and collective activity.[23] However, reli-

21 See N. Chagnon, *Yanomamö: The Fierce People* (New York: Holt, Rinehart & Winston, 1968).

22 See E. Dutton, D. Van der Linden, and R. Lynn, "Population Differences in Androgen Levels: A Test of the Differential-*K* Theory," *Personality and Individual Differences*, 90 (2016): 289–295.

23 A. J. Figueredo, D. Andrzejczak, D. Jones et al., "Reproductive strategy and ethnic conflict: Slow Life History as a protective factor against negative ethnocentrism in two contemporary societies," *Journal of Social,*

giousness is more nuanced than a mere slow LHS because it is associated with "negative ethnocentrism" as well—with hostility to outsiders. It is, thus, an adaptation to a very specific niche and the more *K*-strategy an individual or group is the less strongly associated the different components of *K* become, precisely because of adaptation to a very specific niche, this being the essence of a *K* strategy. In this sense, the GFP will start to break-up.[24] It should be noted that new adaptations can arise relatively quickly and have done, especially over the last 40,000 years, 100 times more quickly than was earlier the case.[25]

Different stages of societal development require different levels of *K*. Pastoralism requires greater impulse control, planning, and cooperation than hunter-gathering, and settled agriculture requires these traits to a greater extent than does pastoralism. The simple fact of adopting pastoralism, and especially agriculture, is almost certain to alter the genotype. It means that you have an ever-larger group of people squeezed into an ever-smaller amount of space.[26] Hunter-Gatherers such as the Yanomamö have been shown to frequently fission; to have arguments and to split up and form separate groups. This is possible because the population density in an enormous rainforest is so low and there is sufficient forest in which to forage for food. Once agriculture has developed, this process of fission, though it remains possible, is much more difficult. If a sub-group cannot get along with the main group, then it is more likely that they will be ostracized and left in a situation where they cannot look after themselves. Ultimately, this would lead to their demise: these kinds of people would be less able to pass on their genes than would be the case among hunter-gatherers. Accordingly, the adoption of agriculture will alter the modal personality of a group in a more

Evolutionary and Cultural Psychology, 5 (2011): 14–31.

24 M. A. Woodley, "The cognitive differentiation-integration effort hypothesis: A synthesis between the fitness indicator and life history models of human intelligence," *Review of General Psychology*. 15 (2011): 228–245.

25 J. Hawks, E. Wang, G. Cochran, and R. Moyzis, "Recent acceleration of human adaptive evolution," *PNAS,* 104 (2007): 20,753–20,758.

26 G. Cochran and H. Harpending, *The 10,000 Year Explosion: How Civilization Accelerated Human Evolution* (New York: Basic Books, 2009).

pro-social direction. It will promote a higher level of GFP, so more intensely removing from the gene pool alleles that would make you low in GFP than would hunter-gathering.

Ecological Harshness and Selection for Modal Personality

In addition, the harsher the ecology, the more intense would be the selection both for a slow Life History Strategy, GFP (or components of it, as the strategy becomes very slow) and, for that matter, intelligence. Consistent with this, there is abundant evidence that the average intelligence of populations, including the prevalence of alleles that are indirectly associated with intelligence, increases as you move away from the equator.[27] The "national IQs" that demonstrate this have been heavily criticised and, thus, recalculated to take these criticisms into account. They correlate with the originals at close to 0.9.[28] National average IQ is associated with all aspects of civilization, so a decline in national average IQ would eventually lead to the decline of civilization, as noted earlier. Similarly, Life History Strategy—including modal personality—broadly becomes slower in ethnic groups the colder is their historical ecology, especially when comparing groups that have all adopted agriculture.[29] Broadly, geographical location appears to predict the modal Life History Strategy in different polities.[30]

Furthermore, as the ecology becomes harsher the level of group-selection intensifies, with the triumphant groups tending to combine high intelligence with high positive and negative ethnocentrism, and thus with high GFP. Between the end of the Medieval Period and the early eighteenth century, the world's

27 D. Piffer, "Correlation between PGS and environmental variables," *RPubs* (2018), https://rpubs.com/Daxide/377423.

28 R. Lynn and D. Becker, *The Intelligence of Nations* (London: Ulster Institute for Social Research, 2019).

29 J. P. Rushton, *Race, Evolution, and Behavior: A Life History Perspective* (New Brunswick, NJ: Transaction, 1995).

30 A. J. Figueredo, T. Cabeza de Baca, H. Fernandes et al., "A Sequential Canonical Cascade Model of Social Biogeography: Plants, Parasites, and People," *Evolutionary Psychological Science*, 3 (2017): 40–61.

climate became increasingly cool, reaching the "Maunder Minimum" at the beginning of the eighteenth century. This would have strongly selected not just for intelligence—to survive in the ever more difficult ecology—but also for Conscientiousness, due to the impulse control and planning involved. The famines and plagues that were widespread in this period disproportionately killed those of lower socioeconomic status; those whom we would predict to be of lower intelligence and low in GFP. For example, the Black Death, in the mid-fourteenth century, killed about 40 per cent of the English population but around 80 per cent of its serfs and free labourers, the Medieval equivalent of the "working class."[31] In addition, this heightened competition set off wars, resulting in intense "group selection" and, thus, selection for the personality traits associated with a slow Life History Strategy.[32]

The apex of intense group selection was found among European peoples around the time of the Maunder Minimum, this being reached approximately half a century before the Industrial Revolution began. It has similarly been found, based on assorted measures, that the period around 1800 was the peak of strongly group-oriented values in these societies, based on in particular on the high use of altruistic words in representative samples of texts.[33] A number of other markers indicate not just increasing intelligence but also increasing empathy across time between the Middle Ages and nineteenth century including decreasing cruelty to animals and decreasingly violent judicial punishments.[34]

31 B. Dodds, "Patterns of decline: Arable production in England, France and Castile, 1370–1450," in B. Dodds and R. Britnell, (Eds.). *Agriculture and Rural Society After the Black Death: Common Themes and Regional Variations* (Hatfield: University of Hertfordshire Press, 2008).

32 M. A. Woodley and A. J. Figueredo, *Historical Variability in Heritable General Intelligence: It's Evolutionary Origins and Sociocultural Consequences* (Buckingham: University of Buckingham Press, 2013).

33 S. C. Hertler, A. J. Figueredo and M. Peñaherrera Aguirre, *Multilevel Selection: Theoretical Foundations, Historical Examples and Empirical Evidence* (London: Palgrave MacMillan, 2020).

34 See, E. Dutton and M. A. Woodley of Menie, *At Our Wits' End: Why We're Becoming Less Intelligent and What It Means for the Future* (Exeter: Imprint Academic, 2018).

Christianity also banned infanticide and contraception, compelling people to only have children if they could be sure they could afford them.[35] This caused them, in turn, to delay marriage; selecting out those who were poorer and thus, on average, less intelligent and lower in Conscientiousness. Put simply, England, for example, at the dawn of the Industrial Revolution was likely the highest in pro-social personality that it had ever been, due to the unforgiving selection pressures to which it had long been subjected.

Niches and Personality

However, in understanding selection for personality there is a slight complication that we need to get to grips with. All else being equal, intelligence is, up to a very high level at least, inherently "good" in terms of survival in all environments. This is not the case with "Conscientiousness" or "Agreeableness" as, in some environments, you need to be impulsive or aggressive. As a society develops and becomes more complex, there will be more pressure to find niches in which to survive, which is why the *K*-traits become less strongly correlated at very high levels of *K*. This, in turn, means that there will be more variation within the population in terms of personality type. There may be in niches in which you must be very high in Conscientiousness, for example, and others where survival depends on combining high IQ with low Conscientiousness. Most obviously, incrementalist scientists have high Conscientiousness but creative scientists have relatively low Conscientiousness.[36] A complex society requires both types. Consistent with this model, there is less personality variation in less complex societies.[37]

35 G. Meisenberg, *In God's Image: The Natural History of Intelligence and Ethics* (Kibworth, Leics: Book Guild Publishing, 2007).

36 D. K. Simonton, "Varieties of (Scientific) Creativity: A Hierarchical Model of Domain-Specific Disposition, Development, and Achievement," *Perspectives of Psychological Science,* 4 (2009): 441–452; Dutton and Charlton, *The Genius Famine, op. cit.*

37 P. Smaldino, A. Lukaszewski, C. von Rueden, and M. Gurven, "Niche Diversity Can Explain Cross-Cultural Differences in Personality Structure," *Nature Human Behaviour,* 3 (2019): 1276–1283.

In essence, then, personality variation depends on all kinds of micro-niches that have historically developed within complex societies, based on such factors as geography, social class, population density and so forth. These niches would involve gene complexes coming together and breaking apart, often relatively quickly.[38] In addition, under conditions of group-selection, the group which operated the most efficiently—due to having the precisely optimum balance of different micro-niches all cooperating as a kind of super-organism—would displace the one which lacked the correct balance, much as a successful ants' nest requires the correct balance between different castes of ant. This raises important questions with regard to eugenic intervention on personality, to which we will return below.

Selection for Religiousness

Religiosity would, it seems, have been selected in tandem with this process. It has been demonstrated that we became more religious, seemingly for genetic reasons, between the Middle Ages and the Early Modern Era.[39] We know religiousness is subject to evolutionary selection because, like all such traits, it is a human universal, it correlates with mental and physical health and with longevity, it correlates with fertility, it is heightened at times of stress (implying it is a kind of instinct),[40] there are specific parts of the brain associated with it (especially the posterior medial frontal cortex, which is also associated with ethnocentrism),[41]

38 See, M. A. Woodley of Menie, S. Luoto, M. PeñaherreraAguirre and M. A. Sarraf, "Life History Is a Major Source of Adaptive Individual and Species Differences: A Critical Commentary on Zietsch and Sidari (2020)," *Evolutionary Psychological Science,* 7 (2021): 213–231.

39 E. Dutton and G. Madison, "Execution, Violent Punishment and Selection for Religiousness in Medieval England," *Evolutionary Psychological Science,* 4 (2018): 83–89.

40 R. Vaas, "God, gains and genes," in E. Voland and W. Schiefenhövel, (Eds). *The Biological Evolution of Religious Mind and Behavior* (New York: Springer, 2009).

41 C. Holbrook, K. Izuma, C., Deblieck, C. et al., "Neuromodulation of group prejudice and religious belief," *Social Cognitive and Affective Neuroscience,* 11 (2016): 387–394.

and it is roughly 40 per cent genetic, being as high as 66 per cent genetic on measures of pronounced religiosity such as being "born again."[42]

Those who collectively worship a moral god are more likely to be married, less likely to be divorced and more likely to have children.[43] They have lower rates of genetic physical and mental illness, live longer, are more likely to get over serious illnesses, and have lower morbidity.[44] Religiousness is, therefore, an evolved *cognitive bias*. It's the evolutionary "norm." Atheism, and other deviations from traditional religious belief, in modern populations, can be understood as a mutation. Consistent with this, atheism is associated with other markers of mutation such as left-handedness, autism, poor physical health, poor mental health, and being physically asymmetrical, as we will see in more detail later.[45] In addition, being "right wing," which is highly relevant because conservatism strongly and robustly correlates with religiousness,[46] is associated with being rated as having a more symmetrical and attractive face: in other words, a lower level of

42 M. Bradshaw and C. Ellison, "Do genetic factors influence religious life? Findings from a behavior genetic analysis of twin siblings," *Journal for the Scientific Study of Religion*, 47 (2008): 529–544.

43 M. Blume, "The reproductive benefits of religious affiliation," in E. Voland and W. Schiefenhövel, (Eds), *The Biological Evolution of Religious Mind and Behavior* (New York: Springer, 2009).

44 H. Koenig, "Religion, Spirituality, and Health: The Research and Clinical Implications" *ISRN Psychiatry* (2012), http://dx.doi.org/10.5402/2012/278730.

45 E. Dutton, G. Madison, and C. Dunkel, "The Mutant Says in His Heart, 'There Is No God': The Rejection of Collective Religiosity Centred Around the Worship of Moral Gods is Associated with High Mutational Load," *Evolutionary Psychological Science*. 3 (2017): 233–244.

46 T. J. Bouchard, "Authoritarianism, Religiousness, and Conservatism: Is "Obedience to Authority" the Explanation for Their Clustering, Universality and Evolution?" in E. Voland and W. Schiefenhövel, (Eds.), *The Biological Evolution of Religious Mind and Behavior* (New York: Springer, 2009).

mutational load.[47] Among men, it is associated with being taller[48] and stronger,[49] these both being indicators of genetic health, as females would select for tall and muscular males.

Among women, gender orientation fantasies and sadistic fantasies are negatively correlated with intrinsic religiousness while masochistic fantasies are negatively associated with fundamentalism.[50] This means that atheism may be associated with deviant sexuality, this tending to reduce fitness. Interestingly, being rated as physically attractive is also associated with accepting the principles of evolutionary psychology.[51] This is presumably because, being low in mutational load, such people are more logical—there is a weak correlation between facial symmetry and general intelligence[52]—and less overwhelmed by negative emotions and feelings of resentment which might cause them to try to create a self-esteem enhancing yet empirically inaccurate worldview, such as that evolutionary psychology is "problematic" and that they, being highly moral, must fight it.

How Did Belief in Gods Develop?

Quite how we came to believe in gods is unclear. One possibility is that there is an evolutionary advantage to over-detecting

47 N. Berggren, H. Jordahl and P. Poutvaara, "The Right Look: Conservative Politicians Look Better and Voters Reward It," *Journal of Public Economics*, 146 (2017): 79–86.

48 R. Arunachalam and S. Watson, "Height, Income and Voting," *British Journal of Political Science*, 48 (2018): 1027–1051.

49 M. B. Petersen, D. Sznycer, A. Sell et al., "The Ancestral Logic of Politics: Upper Body Strength Regulates Men's Assertion of Self-Interest Over Economic Redistribution," *Psychological Science*, 24 (2013): 1098–1103.

50 T. Ahrold, M. Farmer, P. Trapnell, and C. Meston, "The Relationship Among Sexual Attitudes, Sexual Fantasy, and Religiosity," *Archives of Sexual Behavior*, 40 (2010): 619–630.

51 A. Ward, T. English, and M. Chin, "Physical attractiveness predicts endorsement of specific evolutionary psychology principles," *PLoS ONE* 16(2021): e0254725.

52 G. Banks, J. Batchelor, and M. Mcdaniel, "Smarter people are (a bit) more symmetrical: A meta-analysis of the relationship between intelligence and fluctuating asymmetry," *Intelligence*, 38 (2010): 393–401.

agency. When you hear a sound in the forest and you assume it's a wolf then if you are wrong you have lost nothing and if you are right then this will have saved your life, so there is an advantage to over-detecting agency and so perceiving evidence of it everywhere, even behind the world itself.[53] Human altruism may also be relevant to our development of belief in gods and spirits. A key dimension of altruism is empathy, the ability to infer mental states from external signals and thus correctly intuit the feelings of others. Those who are extremely high in this trait, such as schizophrenics, are particularly prone to perceiving signals of agency everywhere, even in the world itself. Those who believe in God, but are psychologically normal, also have this inclination, though to a less pronounced degree and, unsurprisingly, stereotypical high functioning autistics who are very low in empathy, tend to be atheists.[54] Autism is characterized by low empathy (in the sense of theory of mind), high sensitivity to stimuli (including one's own feelings), obsessiveness and a strong desire for order, structure, and systematizing, leading to anxiety about chaos and thus a need for certainty.[55]

Once people believe in God, they are also at an advantage because they will be more moral (more cooperative) and less stressed, and so less likely to get cast out by the band.[56] Religiousness will also be sexually selected for because, in a slow LHS ecology, it tells the female that the male will not abandon her and it tells the male that the female will not cuckold him, because these religions tend to promote "patriarchy" as the will of God. Patriarchy—the male control of female sexuality—is adaptive because if males can trust that their females will not cuckold them, then the males will get into fewer fights and the society will be more internally cooperative.[57]

53 J. Barrett, *Why Would Anyone Believe in God?* (Lanham, MD: AltaMira Press, 2004).

54 See, R. Dunbar, *How Religion Evolves and Why It Endures* (London: Penguin, 2022), 120–123.

55 Baron-Cohen, "The extreme male brain theory of autism," *op. cit.*

56 A. Norenzayan and A. Shariff, "The origin and evolution of religious pro-sociality," *Science*, 322 (2008): 58–62.

57 See, E. Dutton and J. O. A. Rayner-Hilles, "Early Modern Witches

Religion as a Selection Pressure

So, religiousness will be selected for and the group that is more religious will be selected for. Intelligence correlates with impulse control, rule-following, forethought, altruism, and cooperativeness. It also correlates with a propensity for openness to decreasingly instinctive ideas, such as a single moral God, which less advanced human groups don't believe in.[58] As religiousness is selected for, both intelligence and the cooperative personality, involving altruism and high impulse control ("General Factor of Personality," GFP),[59] are simultaneously also selected for, meaning the nature of society changes over time.

However, and this is the central point, religiousness itself then starts to select even further for these very same evolutionarily adaptive traits, often through superficially bizarre or even brutal rituals and rules which, horrible though they can be, are strongly evolutionarily adaptive. And precisely because they are presented as God's will or the behaviour engaged in by the gods, they are more likely to be followed. Once people felt that God was real then they would no longer always be asking themselves, "What do other people expect me to do?" but rather, "What would God want me to do?" And, in general, the kind of behaviour God would be understood to want would be group-oriented, pro-social behaviour.[60] This is because taboos and myths in all religions reflect what has worked successfully in previous generations.

Religious myths are encoded by the evolutionary survivors, because those who believed that God wanted different kinds of

and Demonic Sexual Fantasies: An Evolutionary Perspective," *Mankind Quarterly*, 61 (2021): 936–962.

58 See Dutton, Madison and Dunkel, "The Mutant Says in His Heart, 'There Is No God,' *op. cit.*

59 D. Van der Linden, J. te Nijenhuis and A. Bakker, "The General Factor of Personality: A meta-analysis of Big Five inter-correlations and a criterion related validity study," *Journal of Research in Personality*, 44 (2010): 315–327.

60 See Y. Sela, T. Shackelford, and J. Liddle, "When religion makes it worse: Religiously motivated violence as a sexual selection weapon," in D. Sloane and J. Van Slyke, (Eds.), *The Attraction of Religion: A New Evolutionary Psychology of Religion* (London: Bloomsbury, 2015).

behaviour from that promoted by the myths simply didn't pass on their genes. Religious ideas are products of "evolution" in the sense that the negative "mutations" have been weeded out over generations of selection. In effect, morals and other ideas attributed to God are, in reality, the net experience of human endeavour that has happened to be successful. Presenting that experience in terms of divine laws is, therefore, an extremely powerful way of preserving them to ensure future evolutionarily adaptive behaviour. Reading the mind of God, and empathizing with Him, thus ensures the individual's and the group's evolutionary success. It has been shown that as groups become more complex—and especially once they develop cities—their gods become more concerned with "morality" as religion of this kind forces them to cooperate with relative strangers, so maintaining the ethnocentrism of society. Such a religion is an insurance policy of mutual cooperation.[61]

61 A. Norenzayan, *Big Gods: How Religion Transformed Cooperation and Conflict* (Princeton, NJ: Princeton University Press, 2013).

Six

The Fall of Pro-Social Personality, Mental Health, and Religiosity

Screaming Like a Wounded Animal

On 20th January 2017, Donald Trump (b. 1946) was sworn in as President of the United States in the wake of one of the most fascinating and acrimonious election campaigns in American history. The tycoon and media personality was widely mocked when he initially announced his candidature for the Republican Party, with it being assumed that he would be knocked out early in the "Primaries." After all, the businessman had no political experience whatsoever. Unlike his eventual Democrat opponent—former First Lady, Senator, and Secretary of State Hillary Clinton (b. 1947)—Trump was nowhere close to America's political "Establishment." However, Trump's blunt, populist, nationalistic campaign—in which he made a point of breaking numerous political taboos—propelled him to the Republican Party's candidacy and then to a shock presidential victory, albeit only via the Byzantine workings of the Electoral College, as Trump lost the "popular vote."

Trump dismissed the left-biased media as "fake news" and denigrated its journalists, sometimes even while they were present at his often angry rallies. He insinuated that violence might reasonably be employed against protestors at his charismatic meetings. He was openly nationalistic, proclaiming that he would "Make America Great Again" and claiming that most Mexican immigrants were "drug-dealers" and "rapists" when announcing his party candidacy. He even said, in a presidential debate, that he might put Hilary Clinton on trial for corruption. Trump polarised America—with half of Americans agreeing with him and appreciating his refreshing honesty and half of them fearing and loathing him. Trump won, possibly to his own surprise.

The morning after Trump's triumph, an old friend of mine from America—a registered Democrat living in New York—posted that he and his girlfriend had "cried" when Trump's victory became clear. In actuality, they probably got to the point of being slightly tearful and they did this in the privacy of their own apartment. By contrast, on the day of Trump's inauguration, "Social Justice Warriors," as they were termed at the time, who supported Hillary Clinton and who effectively regarded Trump as wicked, turned up on Pennsylvania Avenue to protest and express their feelings very publicly. One of them, Jessica Starr, fell to her knees and at the exact moment that Trump was announced, via the loudspeakers, to be "President of the United States," emitted what has been described as a "visceral, primal scream." She broke-down crying and sobbed to a journalist from Britain's ITV: "I'm sorry, I love people, this is not America, this is not what we want."[1]

Only a few decades earlier, such behaviour—to undergo an emotional breakdown in public—would have been regarded as socially unacceptable. Most people would have taken the view that such a woman—with her complete lack of impulse control, her absence of emotional regulation, and with the evident intensity of her negative feelings—was mentally ill and should receive medical treatment. This would be perceptive, as over 50 per cent of females aged under-30 who could be categorised as "far left" have been diagnosed with a mental illness, usually depression. This is characterized by feeling "negative feelings" with great intensity, and it becomes increasingly less prevalent as the sample gets older, and also as we move towards the political "right."[2]

1 M. Giessler, "Protester who screamed at Trump's inauguration: 'This is not America,'" *ITV News* (20th January 2017), https://www.itv.com/news/2017-01-20/protester-who-screamed-at-trumps-inauguration-this-is-not-america.

2 E. Kirkegaard, "Mental Illness and the Left," *Mankind Quarterly,* 60

Were we to go back to the 1920s, most people would agree that Jessica Starr should be forcibly removed from society, confined to an asylum until she recovered. They would rightly understand that depression, and other negative feelings, seem to spread like a virus, damaging the mental health of society as a whole.[3] How could a society have become so high in mental illness that Jessica Starr's histrionic outburst was regarded as acceptable, albeit unusual enough to "go viral" online? Certainly, Martin Giessler, the ITV correspondent who interviewed Starr, did not regard her behaviour as a problem, writing merely: "There was so much more I could have asked her, but I left her to have her moment in peace."[4]

The Breakdown of Selection for Pro-Social Personality

We have already discussed the consequences of the Industrial Revolution in terms of the breakdown of selection for intelligence. We would expect it to have a similar impact on personality and religiousness, though the related sequence might be slightly more complicated.

To put it briefly, firstly, there would be a rise in mutational load which would in itself mean more people surviving childhood who deviated from what had been selected for up until 1800. As such, they would be individualistic and so left-wing, atheistic, low in Agreeableness, low in Conscientiousness, and high in Neuroticism. Consistent with this, atheism is, indeed, associated with all of these traits[5] and is, in addition, associated with evidence of physical mutational load, such as being left-handed, as we saw earlier. Atheism is also associated with autism, one characteristic of which is low empathy; that is, in part, poor theory of mind

(2020): 487–510.

3 T. E. Joiner, "Contagious depression: existence, specificity to depressed symptoms, and the role of reassurance seeking," *Journal of Personal and Social Psychology*, 67 (1994): 287–296.

4 Giessler, "Protester who screamed at Trump's inauguration," *op. cit.*

5 J. Gebauer, W. Bleidorn, S. Gosling et al., "Cross-Cultural variations in Big Five relationships with religiosity: A sociocultural motives perspective," *Journal of Personality and Social Psychology*, 107 (2014): 1064–1091.

and, thus, little interest in external signals of internal states.[6] Similarly being "liberal" or "left-wing" correlates, in contrast to being conservative or "right-wing," with low Agreeableness, low Conscientiousness and high Neuroticism.[7]

Liberalism and Anti-Social Personality Traits

Liberalism also correlates with being Machiavellian—and thus selfish and obsessed with personal power—and with the vulnerable Narcissistic Personality, meaning you are deeply insecure (implying high Neuroticism) and thus you crave adoration. You, therefore, create a "false self," that you don't entirely "feel," which is perfect and superior, and you try to persuade yourself that this is your true self. Perhaps you notice what those who have power and prestige do and imitate it in order to reassure yourself that you somehow matter. You can reassure yourself of this, and gain adoration, at least in a broadly liberal society, via left-wing activism, in which you prove your supposed moral superiority to yourself and others.[8] In this regard, challenges to one's masculinity produce an aggressive response in liberal, but not in conservative males, implying a sense of insecurity about power in the former.[9] In a conservative society, we might expect the same psychological type to "conservatism-signal," hence American writer

6 E. Dutton, G. Madison, and C. Dunkel, "The Mutant Says in His Heart, 'There Is No God': The Rejection of Collective Religiosity Centred Around the Worship of Moral Gods is Associated with High Mutational Load," *Evolutionary Psychological Science.* 3 (2017): 233–244.

7 B. Verhulst, P. Hatemi and N. Martin, "Corrigendum to "The nature of the relationship between personality traits and political attitudes" [Personal. Individ. Differ. 49 (2010): 306–316]," *Personality and Individual Differences,* 99 (2016): 378–379; E. Kirkegaard, "Mental Illness and the Left," *op. cit.*

8 J. Moss and P. O'Connor, "The Dark Triad traits predict authoritarian political correctness and alt-right attitudes," *Heliyon*, 6 (2020): e04453.

9 S. H. DiMuccio and E. D. Knowles, "Something to Prove? Manhood Threats Increase Political Aggression Among Liberal Men" *PsyArXiv Preprints* (2022), https://psyarxiv.com/qnpw4/?fbclid=IwAR2E9jEkPFkBIgCSB8Yt9YUkD4EMyza4QxaEwm0CPm9MLTlwrfpiwlSw9x8.

Jim Goad refers to woke females as "the New Church Ladies,"[10] the modern equivalent of the highly conservative spinsters of a century ago.

Interestingly, being on the "Alt Right" correlates with psychopathology.[11] Psychopathology, Narcissism and Machiavellianism compose the "Dark Triad" of anti-social traits.[12] This may be because, in a liberal society, being "Alt Right" is dangerous—with psychopaths being attracted to dangerous situations—and displays very low cooperativeness. It may also be that in order to be very strongly group-oriented you must be very low in feelings of empathy for random individual members of your own group. Congruous with the evidence of vulnerable Narcissism among liberals and its relationship with activism, conservatives find life more inherently meaningful, are more satisfied with life and feel a sense of existential purpose more strongly than do liberals.[13] It would follow that liberals would deal with this void via some kind of mission that they would create for themselves. As Canadian psychologist Jordan Peterson (b. 1962) put it on Australian panel show *Q&A* in February 2019:

> I think, generally, people have things that are more within their personal purview that are more difficult to deal with and that they're avoiding, and that, generally, the way they avoid them is by adopting pseudo-moralistic stances on large-scale social issues so that they look good to their friends and their neighbours.

In response, the young, chubby, female, woke audience member who had questioned him, and was sitting next to a female friend

10 J. Goad, *The New Church Ladies: The Extremely Uptight World of "Social Justice"* (Stone Mountain, GA: Obnoxious Books, 2017).

11 Moss and O'Connor, "The Dark Triad traits predict authoritarian political correctness and alt-right attitudes," *op. cit.*

12 D. Paulhus and K. Williams, "The Dark Triad of Personality: Narcissism, Machiavellianism and Psychopathy," *Journal of Research in Personality*, 36 (2002): 556–563.

13 D. Newman, N. Schwarz, J. Graham, and A. Stone, "Conservatives Report Greater Meaning in Life than Liberals," *Social Psychological and Personality Science*, 10 (2019): 494–503.

who had shaved all her hair off, shook her head and mock-laughed contemptuously.

The correlation between leftism and Machiavellianism may be because, high in negative feelings, such people are paranoid, and thus crave a sense of control and power. This would be augmented in a broadly liberal society, because such people would understand that they could attain power by competitively signalling their liberalism. However, even in a conservative society, some people—extremely high in Neuroticism and thus in paranoia, a burning sense of unfairness, a profound desire for control and an inability to accept others having power over them—would still be leftist, though, as already noted; we would predict that many people with this psychology, albeit in a less pronounced form, would be conservative in these circumstances.

Broadly, left-wing people are individualists who, when they feel mistreated, will turn on their own group, often by collaborating with outsiders. By contrast, when conservatives feel unfairly treated, they turn on members of other groups. Turning on your own group—being a potential traitor—means you will not care about the good of your group, only about yourself and the degree to which you have power. It thus makes sense that leftists tend to identify with the out-group. The moral circle of the conservative is kin and ethnic group; the moral circle of the liberal is much more distant from self; allowing collaboration in the individual interest.[14]

Liberalism and Moral Foundations

Humans are concerned with five key moral foundations: In-group loyalty, Obedience to Authority, Sanctity, Harm Avoidance, and Equality. These reflect the fact that humans are pack animals, rendering them concerned with the first three, but they are also individuals who must try to ascend the hierarchy and survive, causing them to be concerned with the latter two. The first three are group-oriented "Binding" values, while the latter three are "Individualistic" values, where you are concerned about your

14 A. Waytz, R. Iyer, L. Young et al., "Ideological differences in the expanse of the moral circle," *Nature Communications*, 10 (2019): 1–12.

own safety, that you have as least as much as everyone else, and that you are not dominated by others. Conservatives are concerned with all five of these foundations. Liberals are only concerned with individualizing foundations. Thus, conservatives are group-selected, while liberals are individually-selected—though they may pretend to care about others, specifically disadvantaged out-groups, as means of appealing to a general desire for "equality" and "harm avoidance" which even conservatives hold. However, conservatives also value in-group loyalty, obedience to authority and sanctity.[15]

The liberal appeals to these "individualistic" moral foundations because he is individualistic, and so these are the ones he feels strongly about; he feels that he deserves more relative power ("equality" or "fairness") and he feels others have power over him ("harm"). He is a fast Life History Strategist, not evolved to be part of a group, so he is high in mental instability, meaning he feels disempowered and identifies with disempowered groups, and this gives him an incentive to desire power and, high in negative feelings, he distrusts those who are in power over him. Signalling interest in individualistic values is useful as even conservatives care about these, though alongside the others. Moreover, perceiving himself as weak, the liberal will have to play for status covertly, and this is precisely what virtue-signalling about "equality" and "harm avoidance" allows him to do. Hence, females, being physically weaker, tend to use this tactic to a greater extent than males,[16] and liberal males become more conservative, and more overtly confrontational, when administered testosterone.[17] This covert method of virtue-signalling permits the liberal to attain personal power, despite it having been shown that he is

15 J. Graham, J. Haidt and B. Nosek, "Liberals and Conservatives Rely on Different Sets of Moral Foundations," *Personality Processes and Individual Differences,* 96 (2009): 1029–1046.

16 J. F. Benenson, "The development of human female competition: allies and adversaries," *Philosophical Transactions of the Royal Society B,* 368 (2013): 2013007920130079.

17 R. S. Alogaily, "Testosterone Administration Induces a Red Shift in Democrats," in *Essays in Behavioral Economics and Neuroeconomics Economics* (PhD Thesis: Claremont Graduate University, CA, 2022).

motivated primarily by envy, not by compassion, with "malicious envy" being the single biggest motivator of radical liberals.[18] Self-deception helps in covertly attaining power. If you genuinely believe, as hard as you can, in "equality", and you genuinely believe that you're motivated solely by helping the marginalized, you will not so easily be seen as a competitor. Women tend to genuinely believe that they are not competing with each other. In reality, they are competing all the time, though this cannot be known because female social groups tend to strongly prize "equality" and seeming "nice." Only if this tactic fails do they start acting to have their female competitor socially excluded.[19] But they are still "nice," even if they do this: the excluded female has been, somehow, "mean."[20]

In addition, the liberal will "victim signal"—another correlate of Narcissism and Machiavellianism—as a covert means of attaining power through eliciting sympathy.[21] In a sense, the liberal "doth protest too much." He loudly proclaims that he is compassionate and kind in order to hide the fact that he is selfish and ruthless and in order to allay his own feelings of inferiority by convincing himself and others of his moral superiority. Unsurprisingly, signalling your morality predicts moral hypocrisy.[22] The conservative is actually compassionate and kind, so has no need to proclaim this and, being happier in himself and with the world, has no need to act to allay feelings of inferiority. It is rather like an insecure middle-class person—such as Hyacinth Bucket ("pronounced 'Bouquet'") of the early-1990s sitcom *Keeping Up*

18 C. Lin and T. C. Bates, "Each is to count for one and none for more than one: Predictors of support for economic redistribution," (2021), https://doi.org/10.31234/osf.io/3jq4c.

19 J. Benenson, *Warriors and Worriers: The Survival of the Sexes* (Oxford: Oxford University Press, 2014), 177.

20 Benenson, *Warriors and Worriers, op. cit.*, 195.

21 E. Ok, Y. Qian, B. Strejcek and K. Aquino, "Signaling Virtuous Victimhood as Indicators of Dark Triad Personalities," *Journal of Personality and Social Psychology*, 120 (2021): 1634–1661.

22 M. Dong, T. Kupfer, S. Yuan and J.-W. van Prooijen, "Being good to look good: Self-reported moral character predicts moral double standards among reputation-seeking individuals," *British Journal of Psychology*, (2022), doi.org/10.1111/bjop.12608

Appearances—ending-up with a posher accent than a genuinely upper-class person; albeit a disingenuous one.

In that the "liberal" strongly deviates from what would have previously been selected for, we would expect these mental mutations to be paralleled by physical mutations. As already noted, 84 per cent of our genes are expressed in the brain, rendering the brain an enormous target for mutation.[23] Most of these genes are also expressed outside the brain, having effects on many different bodily functions. Therefore, when mutational load is sufficiently high to have impacted the brain, it will surely have impacted the body. As noted earlier, a number of studies have found that conservatives tend to be perceived as more physically attractive than liberals; their faces are more symmetrical, implying lower mutational load.[24] From this difference in attractiveness therefore, we should expect an even larger difference in mental health, and many studies have demonstrated this to be so.[25]

The Break-up of the Religious Bundle

Put simply, the weakening of selection pressures would be expected to lead to a rise in liberalism and atheism. It would also lead to a rise in all other mental disorders, such as depression, bipolar disorder, and schizophrenia, as well as in personality disorders such as Narcissistic Personality Disorder or Borderline Personality Disorder, which we will examine later. However, we might also expect it to be slightly more complicated than this. Religiosity—in the sense of the collective worship of a moral god—has been selected for and it is, in itself, a "bundle" of evolutionarily adaptive mechanisms, such as over-detecting agency

23 S. Negi and C. Guda, "Global gene expression profiling of healthy human brain and its application in studying neurological disorders," *Scientific Reports*, 18 (2017): 897.

24 N. Berggren, H. Jordahl and P. Poutvaara, "The Right Look: Conservative Politicians Look Better and Voters Reward It," *Journal of Public Economics*, 146 (2017): 79–86; R. Peterson and C. Palmer, "Effects of physical attractiveness on political beliefs," *Politics and Life Sciences*, 36 (2017): 3–16.

25 See, E. Kirkegaard, "Mental Illness and the Left," *Mankind Quarterly*, 60 (2020): 487–510.

(meaning you're less likely to get killed in the wilderness because you mistake the sound of an animal moving for merely the wind blowing), obedience to authority, over-detection of causation (causing you to be fascinated by how things work), fervent belief, including in fate (giving you a sense of anxiety-reducing certainty), perceiving the out-group as evil (making you more positively and negatively ethnocentric) and so on.[26]

We have already seen that this form of religiosity substantially developed in the context of complex societies which had themselves manifested due to previous intense selection pressures for intelligence and pro-social personality. So, with the weakening of these pressures, we would expect this "religion bundle" to come apart in all directions and for these "replacement religions"—where there are aspects of the bundle but no collective worship of a moral god—to be correlated with poor mental and poor physical health. Consistent with this, belief in "conspiracy theories"[27]—where an explanation that does not involve agency is more plausible—correlates with depression, anxiety and aspects of psychopathology, via paranoia.[28] Belief in the paranormal is associated with schizophrenia.[29] Extreme liberalism—the "woke" New Religion which is fervent, perceives dissenters as evil, perceives right-wing plots everywhere, and is certain extreme liberals will triumph—is associated with serious depression,[30] as is becoming involved in "New Religious Movements" which tend to deviate from traditional religious norms.[31]

26 See, P. Boyer, *Religion Explained: The Human Instincts That Fashion Gods, Spirits and Ancestors* (London: Heinnemann, 2001).

27 I appreciate that this term is becoming a kind of appeal to insult and emotion; a way of shutting down ideologically inconvenient discussion, rather like the word "racist." Therefore, we must be very clear what we mean by it.

28 A. Goreis and M. Voracek, "A Systematic Review and Meta-Analysis of Psychological Research on Conspiracy Beliefs: Field Characteristics, Measurement Instruments, and Associations with Personality Traits," *Frontiers in Psychology* (2019).

29 N. Dagnall, K. Drinkwater, A. Parker et al., "Conspiracy Theory and Cognitive Style: A Worldview," *Frontiers in Psychology*, 6 (2015): 206.

30 Kirkegaard, "Mental Illness and the Left," *op. cit.*

31 P. Hills, L. Francis, M. Argyle and C. Jackson, "Primary personality

Having looked at the consequences of mutational load, secondly, we would expect the two extremes of the Life History Strategy that we explored earlier to also break up. This is because one is an adaptation to an unpredictable yet easy environment while the other is an adaptation to a predictable yet harsh environment. High civilization, however, creates a predictable *and* easy environment. This would potentially mean selection for the extremes. On the one hand, the rise of contraception means that those who do not want children can avoid having them. Children, and especially large families, would often occur by accident, meaning that they would be associated with impulsiveness and thus with low Conscientiousness. Consistent with this, Conscientiousness is now negatively associated with fertility in Western samples.[32] ADHD, a significant component of which is low impulse control, is thus positively associated with fertility.[33] Extraversion—because it involves risk-taking in pursuit of the strong "positive feelings" gained as a reward—is also associated with fertility. This is because, bluntly, those high in Extraversion will attain greater enjoyment from sex and will, thus, take greater risks in pursuit of it.[34]

Finally, there are people who want children for selfish reasons, such as to attain higher welfare payments, with them spending the extra child-support on luxuries for themselves. Unemployed mothers seem to increase their fertility in line with increased welfare payments.[35] Others may simply be status-seekers who regard a child as an expression of their health; a child is a reason for celebration in most cultures. But, whatever the reason, people

trait correlates of religious practice and orientation," *Personality and Individual Differences,* 36 (2004): 61–73.

32 M. Jokela, "Birth-cohort effects in the association between personality and fertility," *Psychological Science,* 23 (2012): 835–841.

33 N. Mullins, A. Ingason, H. Porter et al., "Reproductive fitness and genetic risk of psychiatric disorders in the general population," *Nature Communications,* 8 (2017): 1–6.

34 M. Jokela, "Birth-cohort effects in the association between personality and fertility," *op. cit.*

35 See, A. Perkins, *The Welfare Trait: How State Benefits Affect Personality* (London: Palgrave Macmillan, 2016).

with the very specific trait of being "incentive-seeking and selfish" have more children than controls.[36] In this sense, we would expect a rise in fast LHS traits, and this would be combined with declining intelligence.

On the other hand, the collapse of child mortality down to negligible levels would mean that those who desired to have children would tend to have them. One predictor of desiring children is Agreeableness, because such people have a strong desire to nurture and to care for others. This trait is now positively associated with fertility.[37] Neuroticism is negatively associated with fertility. People who suffer from depression regard life as pointless and the world as a nefarious place, resulting in them not wanting to have children and resulting in them focusing on their own survival.[38] Consistent with this, Neuroticism is negatively associated with fertility, meaning that, in a sense, we are slowly selecting for mental stability or at least for certain forms of mental stability.[39] This may seem to contradict the expectation that rising mutational load will result in rising levels of genetically-mediated mental instability. But it makes sense: Rising mutational load results in a growing percentage of mentally unstable people with maladaptive inclinations but where these lead to them not desiring to breed then the relatively high presence of such people can only be transient.

The other groups, unsurprisingly, who desire to have children, and thus have them, are those who are traditionally religious and those who are conservative.[40] In this sense, it can be argued that a

36 F. Gutiérrez, M. Gárriz, J. M. Peri et al., "Fitness costs and benefits of personality disorder traits," *Evolution and Human Behavior,* 34 (2013): 41–48.

37 M. Jokela, "Birth-cohort effects in the association between personality and fertility," *op cit.*

38 P. Schönegger, "What's up with anti-natalists? An observational study on the relationship between dark triad personality traits and anti-natalist views," *Philosophical Psychology*, 35 (2022): 1.

39 M. Jokela, "Birth-cohort effects in the association between personality and fertility," *op. cit.*

40 M. Fieder and S. Huber, "Political Attitude and Fertility: Is There a Selection for the Political Extreme?" *Frontiers in Psychology* (2018), https://doi.org/10.3389/fpsyg.2018.02343.

small "remnant population" from pre-Industrial times can gradually be expected to grow as a percentage of the population due to the limited fertility, relative to them, of the mass of the population who came into being mainly due to the period of delay in reducing family size which occurs during the "demographic transition."

The Rise of Genetic Mental Disorders

If dysgenics were occurring, we would expect to see a rise in highly genetically-mediated mental disorders. In addition, this could not be put down entirely to other factors such as to rising paternal age (which is the main cause of new mutations), environmental pollutants, or maladaptive responses to "evolutionary mismatch." The best example of this is unhealthy eating under conditions where tasty food is abundant—humans being evolved to eat such food wherever and whenever they find them under the conditions of resource scarcity to which they are evolved. The rise of mutations over the last century has been on the order of 1 per cent per generation, meaning a 4 per cent increase in mutational load over this period. As we will see in the next chapter, this has had serious physical consequences, including a rise in genetic poor health and a fall in male sperm counts, with there being very little evidence that chemical pollutants are behind this.[41] It has, moreover, been shown that rising paternal age alone is insufficient to account for the extent of the accumulation of mutations over time.[42] A simulation has been produced, using data from Iceland, showing that if selection were intensified to the levels that were prevalent for most of human history, then there would be no rise in mutational load across time.[43] Thus, it is clear that there is a general "fitness factor" and we have be-

41 M. Sarraf, M. A. Woodley of Menie and C. Feltham, *Modernity and Cultural Decline: A Biobehavioral Perspective* (London: Palgrave Macmillan, 2019), 198.

42 Sarraf, Woodley of Menie and Feltham, *Modernity and Cultural Decline, op. cit.*, 200.

43 Sarraf, Woodley of Menie and Feltham, *Modernity and Cultural Decline, op. cit.*, 204.

come less fit across time due to increasing mutational load underpinned by weakened selection pressure.

Congruous with this, there is evidence of a significant rise in psychopathology. This is characterised by anti-social behaviour, impaired empathy, extreme selfishness, and lack of inhibition, among other traits. Clearly, in a highly pro-social species such as humans, we would expect such tendencies to be strongly selected out. It has been argued that the widespread use of capital punishment in Medieval and Early Modern Europe—which disproportionately killed young men who had not yet fathered children—helped to achieve this by killing about 2 per cent of the male population each generation. It helped to tame the European personality, by removing psychopaths from the gene pool.[44] In doing so, it has been shown that it also indirectly selected for religiousness.[45]

In the US, between 1938 and 2007, the percentage of the population that scored above the cut-off point for psychopathic personality disorder, when using a specific instrument to measure it, increased five-fold. In 1938, between 1 per cent and 5 per cent of American college students scored above the cut off point for the various measures of psychopathology. By 2007, making adjustments for the increase in college participation and thus in the range, 40 per cent scored above the cut-off point for "hypomania," an abnormally elevated mood which is itself one of the symptoms of bipolar disorder and is associated with grandiosity and even with hyper-sexuality. Six per cent qualified for depression (approximately a seven-fold increase), 23 per cent for schizophrenia (roughly a six-fold increase), and there were increases of a similar order for other measures of psychopathology. American psychologist Jean Twenge and her team, who conducted the analysis, argue that increasing materialism and decreasing

44 P. Frost and H. Harpending, "Western Europe, State Formation, and Genetic Pacification," *Evolutionary Psychology* (2015): https://doi.org/10.1177/147470491501300114.

45 E. Dutton and G. Madison, "Execution, Violent Punishment and Selection for Religiousness in Medieval England," *Evolutionary Psychological Science*, 4 (2018): 83–89.

religiosity across this period may be a partial explanation.[46]

Alternatively, according to American political scientist Ronald Inglehart (1934–2021), the last half century has been characterized by a decline of materialist values and rise of post-materialist values: Self-expression instead of concerns about security and material wants. He attributes this to rising prosperity.[47] Once a certain level of material comfort is secured, people stop being concerned about this and they move on to the next level of American psychologist Abraham Maslow's (1908–1980) "hierarchy of needs,"[48] which may simply be a desire for power, status and, related to these, self-expression.[49] If you are struggling for food, after all, you are not too concerned about strangers respecting your unusual lifestyle choices. Just look at politics. Socialism with its emphasis on the material plight of the working class has disappeared. Instead, the Western Left is concerned about everybody "feeling" that they are of somehow of "equal" worth; but this can never be achieved, as some people will be better able to accrue wealth and status than others for partly genetic reasons. Post-materialism creates unrealistic expectations and disappointment and, indeed, is a recipe for social discord. It has a similar impact to "elite over-promotion" in which there are too many people with degrees, and who are thus aspirant elites, but only a limited number of "elite" positions, such as membership of legislatures or university positions. The result is intense intra-elite competition which can destabilise society.[50] Lack of religiosity reduces mental-health-elevating activities such as be-

46 J. Twenge, B. Gentile, C. N. DeWall et al., "Birth cohort increases in psychopathology among young Americans, 1938–2007: A cross-temporal meta-analysis of the MMPI," *Clinical Psychology Review*, 30 (2010): 145–154.

47 See R. Inglehart, *The Silent Revolution: Changing Values and Political Styles Among Western Publics* (Princeton, NJ: Princeton University Press, 2015).

48 A. H. Maslow, "A Theory of Human Motivation" *Psychological Review*, 50 (1943): 370–396.

49 See Inglehart, *The Silent Revolution, op. cit.*

50 See P. Turchin, *Ages of Discord: A Structural-Demographic Analysis of American History* (Mansfield, CT: Beresta Books, 2016).

lief in God, prayer, and collective worship.[51]

In light of the research discussed above on declining fitness, it is extremely probable that there is a significant genetic dimension to these increases. The heritability of depression is approximately 0.5, meaning that it is about 50 per cent a product of genes.[52] The heritability of psychopathic personality is also about 0.5.[53] The heritability of schizophrenia is far higher, at about 0.8, meaning that it is overwhelmingly a reflection of genetics.[54] Accordingly, it is likely that there is a genetic component to the rise across time of assorted mental disorders among Americans, and this is particularly true with regard to schizophrenia. In addition, we have already noted that we are selecting for ADHD, as it is associated with fertility.[55] ADHD is close to 0.9 genetic in some studies, with a conservative estimate of 0.74, so its rise can be confidently said to be for genetic reasons.[56] Diagnoses of ADHD in the US increased by 60 per cent between 2004 and 2013, though such a large increase, in comparison to a 10 per cent rise in other disorders, perhaps partly betokens greater awareness of the condition.[57]

On this kind of timescale, however, it seems probable that the genetic effects are indirect, in the sense of rising mutational load precipitating a sub-optimal overall functioning or bringing about a society which severely sub-optimally socialises its off-

51 Twenge, Gentile, DeWall et al., "Birth cohort increases in psychopathology among young Americans, 1938–2007," *op. cit.*

52 F. Lohoff, "Overview of the Genetics of Major Depressive Disorder," *Current Psychiatry Reports,* 12 (2010): 539–546.

53 S. Dhanani, V. Kumari, B. Puri et al., "A systematic review of the heritability of specific psychopathic traits using Hare's two-factor model of psychopathy," *CNS Spectrums,* 23 (2018): 29–38.

54 S. Trifu, B. Kohn, A. Vlasie, and B.-E. Patrichi, "Genetics of schizophrenia (Review)," *Experimental and Therapeutic Medicine,* 20: 3462–3468.

55 N. Mullins, A. Ingason, H. Porter et al., "Reproductive fitness and genetic risk of psychiatric disorders in the general population," *Nature Communications,* 8 (2017): 1–6.

56 S. Faraone and H. Larsson, "Genetics of attention deficit hyperactivity disorder," *Molecular Psychiatry,* 24 (2018): 562–575.

57 T. F. Boat and J. T. Wu, *Mental Disorders Among Low Income Children* (Washington, DC: National Academies Press, 2015), Ch. 12.

spring. It should also be emphasised that there are a variety of pathways via which such conditions could become more prevalent. The first is simply due to rising mutational load, as we have already explored. However, rising mutational load will make the entire organism function less efficiently, meaning that it will employ more of its bio-energetic resources to fight off disease and it will potentially create a sub-optimal environment for its gametes or for any foetus. This is known as "developmental instability." As already discussed, we have witnessed an increase, across time, in left-handedness. This is an acknowledged marker of developmental instability. It implies that the brain has developed asymmetrically, to a greater extent that its "normal" level of slight asymmetry, with relative symmetry being a sign of low mutational load, due to the fact that we are supposed to be symmetrical. Thus, failure to achieve a symmetrical phenotype in the face of disease, or simply inheriting an asymmetrical trait, implies high mutational load. Left-handedness is associated with many mental disorders because these are all indications of developmental instability.[58]

Negative Social Epistasis

Another factor that is downstream of genetics is "Negative Social Epistasis." In essence, humans are evolved to be in a highly social society of other humans who are mentally healthy and who are group-oriented. Humans are extremely environmentally sensitive—they are, after all, born helpless; they are slow Life History Strategists, with such people being extremely environmentally sensitive.[59] Thus, they rely on this society—including its sub-components such as the family—when they are children and throughout their lives, to guide them along an evolutionarily adaptive roadmap of life.

58 M. A. Woodley of Menie, H. Fernandes, S. Kanazawa, and E. Dutton, "Sinistrality is associated with (slightly) lower general intelligence: A data synthesis and consideration of secular trend data in handedness," *HOMO: Journal of Comparative Human Biology*, 69 (2018): 118–126.

59 C. Kuzawa and J. Bragg, "Plasticity in Human Life History Strategy: Implications for Contemporary Human Variation and the Evolution of Genus *Homo*," *Current Anthropology*, 53 (2012): 369–383.

As mutational load rises, they will increasingly come into contact with mentally unhealthy people, such as congenital sufferers from depression. This will cause a sense of depression throughout society, because there is a strong environmental element to depression and it has been shown to be contagious.[60] These people tend to think in maladaptive ways: they promote the view that life is pointless, they promote atheism and the view that life has no meaning, and, being high in negative feelings, they attack traditional and adaptive societal institutions, such as organized religion, which has been shown to be adaptive.[61] In effect, they indirectly help to increase mental illness in society, promoting nihilism and materialistic ways of thinking. Once around 20 per cent of a population question its dominant value system, a "tipping point" is reached and people start to migrate, very quickly, over to what they see as the up-and-coming dominant system.[62] This is due to the social benefits involved in joining the "winning team" at the right time, slightly ahead of the rest of the herd.[63] The result can be seemingly very dramatic changes, such as the sudden collapse of Communism in much of Eastern Europe in 1989 or the sudden collapse of the American-backed government in Afghanistan in 2021.

The consequence of this is that the value system of society will shift dramatically. Those who are "conservative" are equally concerned about all five of the moral foundations that we discussed earlier. Those who are "liberal" are concerned about individualistic foundations and relatively unconcerned about "binding" foundations. Such people signal that they are concerned about equality or about harm to others, but the evidence is clear that this is simply false.[64] They identify with those whom they see

60 Joiner, "Contagious depression," *op. cit.*

61 Kirkegaard, "Mental Illness and the Left," *op. cit.*

62 D. Centola, J. Becker, D. Brackbill, and A. Baronchelli, "Experimental evidence for tipping points in social convention," *Science*, 360 (2018): 1116–1119.

63 M. A. Woodley of Menie and C. Dunkel, "Beyond the Cultural Mediation Hypothesis: A reply to Dutton (2013)," *Intelligence*, 49 (2015): 186–191.

64 J. Graham, J. Haidt, and B. Nosek, "Liberals and Conservatives Rely

as low in power, as they perceive themselves as such, and they will collaborate in their own interests against their own group,[65] always seeing themselves as low in power. This makes them power-hungry and successful in attaining power which, in pre-history especially, meant survival and procreation.

If such people attain cultural power—which they are likely to because the conservatives can sympathise with them but not the other way around—then you will get, for many people, an evolutionary mismatch in which binding foundations are increasingly ignored. This will create dysphoria and depression. Furthermore, group-oriented institutions will be undermined, as will traditional practices of raising children, and preparing them for a harsh yet stable ecology, in favour of "equality" and "harm avoidance." Indeed, people will play for status in a kind of runaway individualism where every adaptive idea or practice is attacked as lacking in "equality." We can expect the result to be a rise in depression, and left-wing thinking is strongly associated with depression,[66] and with not wanting children.[67] In this sense, we can talk of the rise of "spiteful mutants" who push society and everyone in it in a maladaptive direction.[68] This would, in part, reflect the fact that group-selection would have suppressed individualists both genetically and environmentally. With its breakdown, they can increase for genetic reasons, and then for environmental reasons as a tipping point is reached, and they can undermine the societal practices and institutions that would previously have suppressed them.

In my book *Witches, Feminism and the Fall of the West,* I have

on Different Sets of Moral Foundations," *Personality Processes and Individual Differences,* 96 (2009): 1029–1046.

65 Waytz, Iyer, Young et al., "Ideological differences in the expanse of the moral circle," *op. cit.*

66 Kirkegaard, "Mental Illness and the Left," *op. cit.*

67 P. Schönegger, "What's up with anti-natalists? An observational study on the relationship between dark triad personality traits and anti-natalist views," *Philosophical Psychology*, 35 (2022): 1

68 M. A. Woodley of Menie, M. Sarraf, R. Pestow, and H. Fernandes, "Social Epistasis Amplifies the Fitness Costs of Deleterious Mutations, Engendering Rapid Fitness Decline Among Modernized Populations," *Evolutionary Psychological Science*, 3 (2017): 181–191.

looked at the way in which Early Modern witches effectively undermined the degree to which populations were cooperative. They did this by undermining the religiously-sanctioned system of patriarchy. This system was important because if female sexuality (and behaviour in general) was controlled, males would not fear being cuckolded, there would be less inter-male conflict, there would be more inter-male cooperation and it is this "positive ethnocentrism" which is extremely important to triumphing in the battle of group selection. Therefore, purging witches—which was heightened at times of intensified environmental harshness and thus of stronger group selection—was a means of elevating societal fitness. I also noted that many of these witches did indeed practice, and freely admitted to practicing, a form of "folk magic" that undermined the religious monopoly of the patriarchy-promoting Church.[69]

Evolutionary Mismatch and Dysphoria

Under such a system, we are increasingly in an extreme "evolutionary mismatch." Our "evolutionary match" is to live in a small, sparsely populated community, of about 150 people, all of whom are relatively closely related to us—likely distant cousins. The community will be patriarchal, as we are evolved to this,[70] physically and mentally healthy, and will engage in the collective worship of a moral god. Child mortality, however, will be about 50 per cent, so we will constantly be surrounded by death, with religion—to some extent—helping us to negotiate this. This high child mortality is, of course, necessary to ensure the group's genetic physical and mental health. Any deviation from this will create an evolutionary mismatch and, therefore, dysphoria.[71] To give just one example, the amygdala, the brain's centre of fear, is activated when people are shown photographs of members of

69 E. Dutton, *Witches, Feminism, and the Fall of the West* (Whitefish, MT: Radix, 2021).

70 M. Apostolou, *Sexual Selection Under Parental Choice: The Evolution of Human Mating Behaviour* (Hove: Psychology Press, 2013).

71 See, G. Geher and N. Wedberg, *Positive Evolutionary Psychology: Darwin's Guide to Living a Richer Life* (Oxford: Oxford University Press, 2019).

other races whom they do not know—something that happens so quickly that the conscious areas of the brain are unaware of the response. This does not happen when they are shown photographs of strangers who are members of their own race.[72]

Thus, a multi-racial society is subjecting people to constant stress, consistent with a large body of evidence that we tend to socially assort and to mate along the lines of genetic similarity, including along the lines of race. The latter is known as "assortative mating."[73] This stress—especially among environmentally sensitive people who are already prone to stress—could express itself in numerous unpleasant ways, including mass shootings. The majority of American mass-shooters who fail to kill themselves during the incident have been found to be schizophrenic, with very few not having some kind of psychiatric condition,[74] so rising mutational load is also likely to be relevant to the "mass shooter" phenomenon, combined with greater sensitivity to stress among those with psychiatric problems.

72 E. O. Wilson, *The Social Conquest of Earth* (New York: W. W. Norton, 2012), 100.

73 See, E. Dutton, *Race Differences in Ethnocentrism* (London: Arktos, 2019). Races are defined as human groups (sub-species) that differ in genotypic frequencies by virtue of these being adaptations to different ecologies that have long been separated. Geneticists have identified twelve genetic clusters, paralleling the twelve races of Classical Anthropology. Since that separation, in only the last few thousand years, there has been inter-racial contact, creating clines, but the clusters are still clear. When genetic loci are used that actually differ between races (because they reflect ecological adaptation)—rather than those that don't differ much because they don't help you to adapt to an ecology—then 80 per cent of race differences are *between* races. Races differ on numerous important traits, such as aspects of genetic health, meaning that 'race' allows correct predictions to be made; the essence of a scientific category that is a mainly biological construct. See, E. Dutton, *Making Sense of Race* (Whitefish, MT: Washington Summit, 2020).

74 I. Glick, N. Cerfolio, D. Kamis, and M. Laurence, "Domestic Mass Shooters: The Association With Unmedicated and Untreated Psychiatric Illness," *Journal of Clinical Psychopharmacology*, 41 (2021): 366–369.

Why Does Intelligence Predict Not Wanting Children in Advanced Societies?

As we saw earlier, when people are primed with mortality salience, they become more inclined to want to have children.[75] This is likely because such a situation places them in their "evolutionary match" and, thus, reduces dysphoria. We have also already seen that intelligence in modern societies is negatively correlated with fertility. However, it is also associated with simply not wanting to have children at all, among men and women, though more weakly among women.[76] A feasible explanation is that intelligence is somehow related to environmental sensitivity: to plasticity. This would make sense because solving problems—intelligence—involves rising above your instinctive cognitive biases, so people who were less instinctive, and so more environmentally sensitive, would be better at solving problems. This is termed the Intelligence-Mismatch Association Model.[77]

In this regard,

> individuals with high IQ show high environmental influence on IQ into adolescence (resembling younger children), whereas individuals with low IQ show high heritability of IQ in adolescence (resembling adults), a pattern consistent with an extended sensitive period for intellectual development in more-intelligent individuals.[78]

75 I. Fritsche, E. Jonas, Peter Fischer et al., "Mortality Salience and the Desire for Offspring," *Journal of Experimental Social Psychology,* 43 (2007): 753–762.

76 E. Kirkegaard and E. Dutton, "Intelligence and Anti-Natalist Intentions on Dating Sites: An Analysis of the OKCupid Dataset," in R. Lynn, (Ed.), *Intelligence, Race and Sex: Some Controversial Issues: A Tribute to Helmuth Nyborg at 85* (London: Arktos, In Press); S. Kanazawa, *The Intelligence Paradox: Why the Intelligent Choice Isn't Always the Smart One* (Hoboken, NJ: John Wiley & Sons, 2012).

77 See, E. Dutton and D. Van der Linden, "Why is Intelligence Negatively Associated with Religiousness?" *Evolutionary Psychological Science,* 3 (2017): 392–403.

78 A. Brant, Y. Munakata, and D. Boomsma, "The nature and nurture of high IQ: An extended sensitive period for intellectual development,"

Civilization creates an evolutionary mismatch, such as by placing people in extremely dense living conditions with genetic strangers, so inducing stress, and this is felt more keenly by the more intelligent, leading to greater dysphoria and, thus, a greater desire not to have children. This leaves only those who strongly instinctively desire to have children—such as the highly traditionally religious or conservative—or who have them by accident as a consequence of impulsivity or inefficient contraception use. For this reason, civilizations must always collapse, with, perhaps, a remnant of civilization holding out in a small area, as occurred with Byzantium.[79]

An alternative explanation, presented by American researcher Jason Malloy, is that humans have innate drives towards sex and gaining status, with reproduction being mediated through these. For example, in a patriarchal society a woman would attain status by marrying a high-status male and having a child with him and the child would be more likely to survive, and so pass on its own genes, due to its greater access to resources. Due to reliable contraception and female liberation, reproduction is now separated from these drives. If the drive for status were stronger than the drive for reproduction, this would explain why people decreasingly desire children, with higher intelligence predicting better understanding of the negative financial impact of having children.[80] The problem with this model is that it does not explain why highly intelligent males decreasingly want children, nor why they would have wanted them in the first place. It also doesn't explain why, in late Greece and Rome, high status men and women were eschewing breeding in what was still a relatively patriarchal society by modern standards.[81] Thus, I would aver that the Intelligence-Mismatch Association Model is more

Psychological Science, 24 (2013): 1487–1495.

79 See, E. Dutton and J. O. A. Rayner-Hilles, *The Past is a Future Country: The Coming Conservative Demographic Revolution* (Exeter: Imprint Academic, 2022).

80 Jay Man, "A Success Story?" *Jay Man's Blog* (1st October 2012), https://jaymans.wordpress.com/2012/10/01/a-success-story/.

81 J. Glubb, *The Fate of Empires and the Search for Survival* (Edinburgh: William Blackwood & Sons, 1976).

parsimonious.

Kinship and Fertility

The cycles of civilization seem to be predictable and have been compared to the seasons of the year, or the life cycle of an organism. Civilization springs, bears fruit, decays into decadence, and then collapses into a winter of chaos wherein it dies,[82] as we will explore in more detail below.

But there may be an even more fundamental reason for periodic civilizational declension than those outlined above. The most parsimonious explanation must make sense of periodic population collapses in non-human animals as well as in humans. The simplest model is that all animals are evolved to breed with those that are optimally genetically similar to them, in order to avoid the twin problems of inbreeding depression (inheriting harmful double doses of mutant genes) and outbreeding depression. In this context, "outbreeding depression" refers to complex traits or "gene complexes"—that are evolutionarily beneficial and involves many genes cooperating—going awry because the genes are not used to cooperating or are not present. In numerous species, populations grow when they are relatively closely related but go into decline when they become too dense, with random kinship levels being too low.

In parallel with this, fertility is highest between pairs that are optimally related and goes into decline once they become more distantly related. In Iceland, for example, the most fertile couples are third cousins and there is a decline in fertility on either side of this optimum.[83] Interracial couples have higher levels of in-

82 O. Spengler, *The Decline of the West: Volume II: Perspectives of World History* (London: George Allen and Unwin, 1928. Trans. C. F. Atkinson).

83 A. Helgason, S. Pálsson, D. Guðbjartsson and T. Kristjánsson, "An Association Between the Kinship and Fertility of Human Couples," *Science*, 319 (2008): 813–816. However, cousin marriage can develop as a cultural practice in certain ecologies, which do not otherwise strongly select for social trust, in order to ensure strongly cooperative groups via a close level of relatedness. These will dominate other groups in the battle of group selection. In many generations of relatively harsh conditions, mutational load will be low and offspring suffering from inbreeding depression will gener-

fertility than endogamous couples.[84] It has also been found that the female reproductive tract only allows genetically compatible sperm to enter the oocyte, meaning that when a couple suffer from infertility it may be partly because they are insufficiently genetically similar.[85] Studies of fruit fly, voles, mice and many other animals have all noticed the same phenomenon: at a certain point, when random kinship is too low, fertility declines and the population collapses. These fluctuations keep the population optimally genetically healthy and adaptive across time and are the simplest reason for the periodic collapses of human societies, though selection for cooperative personality and other traits can allow some societies to grow larger than others.[86]

Societal collapses, indeed, seem to generally involve a population decline both prior to and after the collapse.[87] This was certainly true of the collapse of Classical Civilization,[88] and, seemingly, Mayan Civilization,[89] though some civilizational de-

ally die young. See, E. Dutton, *Race Differences in Ethnocentrism* (London: Arktos, 2019).

84 Z. Qian and D. Lichter, "Racial Pairings and Fertility: Do Interracial Couples Have Fewer Children?" *Journal of Marriage and Family,* 83 (2021): 961–984; D. Ebeh and S. Jahanfar, "Association between maternal race and the use of assisted reproductive technology in the USA," *SN Comprehensive Clinical Medicine,* 3 (2021): 1106–1114.

85 J. Kekäläinen, "Genetic incompatibility of the reproductive partners: an evolutionary perspective on infertility," *Human Reproduction,* 12 (2021): 3028–3035.

86 R. Fox, "Marry In or Die Out," in A. Maryanski, R. Machalek and J. Turner, (Eds.), *Handbook on Evolution and Society: Toward an Evolutionary Social Science* (London: Routledge, 2015); E. Dutton, and M. L. Herbert, "Why Do Civilizations Collapse? Outbreeding Depression, Mediogamy and the Evolutionary Necessity of the Fall of Advanced Polities." Unpublished Manuscript.

87 Declining fertility, and population decline, was remarked upon during the period prior to the collapse of Classical civilization. See Dutton and Woodley of Menie, *At Our Wits' End, op. cit.,* Ch. 11.

88 J. Tainter, *The Collapse of Complex Societies* (Cambridge: Cambridge University Press, 1988), 140.

89 N. Dunning and T. Beach, "Farms and Forests: Spatial and Temporal Perspectives on Ancient Maya Landscapes," in I. P. Martini and W. Chesworth, (Eds.), *Landscapes and Societies: Selected Cases* (New York:

clensions, such as the Late Bronze Age Collapse, see rapid population decline due to natural disasters and their consequences.[90] In a context of high civilization, however, it would be the more environmentally sensitive who would experience greater dysphoria due to the mismatch of being with insufficiently genetically similar people, meaning the population collapse, or native population collapse, leads to a civilizational collapse.

It might be protested that there appears to be little room for free will in such a system, and that is quite true. As American biologist E. O. Wilson (1929–2021) has noted, if human behaviour is a consequence of an interaction between genetics and environment—which is indeed the case—then there is no space for "free will." It is, however, adaptive to believe that you have free will, for, without this belief, you may become fatalistic and apathetic.[91] Consistent with this suggestion, the feeling that you do not have free will—that is, fatalistic determinism—is associated with suffering from depression.[92]

Sexuality and Sexual Orientation

We would expect this same sequence of evolutionary and environmental processes that lead to an evolutionary mismatch to interfere with the development of sexuality. Bruce Charlton has highlighted the potential impact of mutational load on, for example, sexual development. This would help to explain why the traditional model of acceptable sexual behaviour in Western societies—heterosexual sex within religiously-sanctioned marriage—has broken down in Western countries.

Charlton argues that the optimum development of sexuality involves an extremely complex sequence with ample room for

Springer, 2010). 383.

90 E. Cline, *1177 BC: The Year Civilization Collapsed* (Princeton, NJ: Princeton University Press, 2015).

91 E. O. Wilson, *Consilience: The Unity of Knowledge* (New York: Knopf, 1998), 118–120.

92 J. Fu, Y. Zhao, X. Feng et al., "How is fatalistic determinism linked to depression? The mediating role of self-control and resilience," *Personality and Individual Differences*, 180 (2021): 110992.

things to go wrong at every step of the way. Firstly, there is genetics. Then there are hormone-induced changes, both *in utero*, which would reflect the interaction between the genetics of the mother and the foetus, and in adolescence. There are also "critical learning periods," particularly in childhood, in which there is important imprinting and other psychological development. For example, children would, under Darwinian conditions, have been told that there were "males" and "females" and that it was wrong for "males" to behave like "females." These phases must all be successfully negotiated for sexuality to be adaptive.[93] In this regard, the heritability of male sexuality is around 0.4, while the heritability of female sexuality is around 0.2. It has been proposed that female sexuality is more fluid such that abandoned females, in pre-history, could successfully form family units, bonded by sexual attraction, to co-parent each other's children.[94]

A variety of theories have been suggested for why male homosexuality stays in populations for genetic reasons, such as that its presence reduces inter-male conflict during competition for females which aids group selection, or that it aids the kinship group to have a very small number of members who do not breed and who thus invest their energies in the good of the group. This is especially likely in a context of widespread homophobia, meaning that their homosexuality must be suppressed.[95] One theory for homophobia is that males are evolved to be part of warrior bands in which they will fight for their "band of brothers." An effeminate male, with this correlating with homosexuality, may

93 B. G. Charlton, "Mutation Accumulation and Modern Western Sexuality," *Intelligence, Personality and Genius* (20th July 2015), https://iqpersonalitygenius.blogspot.com/2015/07/mutation-accumulation-and-modern.html.

94 N. Långström, Q. Rahman, E. Carlström, E. and P. Lichtenstein, "Genetic and Environmental Effects on Same-sex Sexual Behavior: A Population Study of Twins in Sweden," *Archives of Sexual Behavior*, 39 (2010): 75–80; A. Jeffrey, T. Shackelford, V. Zeigler-Hill, J. Vonk, and M. McDonald, "The Evolution of Human Female Sexual Orientation," *Evolutionary Psychological Science*, 5 (2019): 71–86.

95 E. Dutton "Why Are Non-Heterosexual Males Attracted to Religious Celibacy? A Case for the 'Gay Shaman' Theory," *Mankind Quarterly*, 59 (2018): 197–215.

be regarded as less interested in so doing and, thus, as a threat to the solidarity of the band.[96] Nevertheless, despite possible evolutionary benefits in very small numbers, homosexuality, bisexuality and transsexuality are all associated with many markers of mutational load, including left-handedness, mental illness and many significantly genetic physical conditions.[97] In this sense, rising levels of sexual deviance would be predicted by weakening selection.

Gender Dysphoria

Gender dysphoria is itself around 0.5 heritable.[98] Further evidence of transsexuality being a manifestation of developmental instability can be seen in the health of transsexuals. Their physical and mental health is far worse than is that of so-called "cisgender" people. Transwomen, compared to normal males, report markedly elevated levels of high cholesterol, high blood pressure, vision problems, hearing problems, chronic pain, arthritis, digestive problems, lung problems, kidney complaints, diabetes, Post-Traumatic Stress Disorder, depression, and anxiety. They are also more likely to abuse alcohol, and even hard drugs such as heroin, and engage in commercial sex work. Transmen are elevated in terms of obesity, smoking, asthma, depression, schizophrenia, sexually transmitted diseases, pubic lice, inflammatory conditions, previous menstrual irregularities, and premature or delayed menarche. They are also elevated in hyperandrogenism and, therefore, in a series of complications associated with elevated testosterone in females, including adrenal hyperplasia and polycystic ovary syndrome.[99]

Congenital adrenal hyperplasia is a very specific defect. It is

96 Benenson, *Warriors and Worriers, op. cit.*, 87.

97 R. Blanchard, "Review and theory of handedness, birth order, and homosexuality in men," *Laterality*, 13 (2008): 51–70.

98 T. Polderman, B. Kreukels, M. Irwig et al., "The Biological Contributions to Gender Identity and Gender Diversity: Bringing Data to the Table," *Behavior Genetics*, 48 (2018): 95–108.

99 S. Reisner, T. Poteat, J. Keatley et al., "Global Health Burden and Needs of Transgender Populations: A Review," *The Lancet*, 388 (2016): 412–436.

a consequence of a genetically caused inability to make cortisol, which leads to overstimulation of the adrenal gland by the pituitary gland. These enlarged, overstimulated adrenal glands overproduce the hormones that they are still able to make, which are mainly androgens.[100] Transmen also have elevated rates of fibromyalgia, though this is not significantly higher in transwomen than it is among the general population.[101] Some of these conditions help to explain why the sufferers are transsexual or reflect the stresses inherent in being transsexual. But others, such as hearing problems, are likely to simply reflect the underlying factor of high mutational load and developmental instability.

Paedophilia has also been found to be associated with developmental instability,[102] with paedophiles being overwhelmingly homosexual in orientation, at a ratio of 2:1.[103] It has been argued that transsexuality—and this could be argued with paedophilia—is a sexual fetish (a paraphilia) in which the individual is sexually aroused by and obsessed with a sexualised object other than a fertile member of the opposite sex, specifically by the idea of themselves as a member of the opposite sex.[104] This would seem to reflect autistic traits, such as obsessiveness and a focus on objects. Unsurprisingly, autism is strongly elevated among transsexuals.[105] Autism can involve an unclear sense of self (due

100 See, P. Hindmarsh and K. Geertsma, *Congenital Adrenal Hyperplasia: A Comprehensive Guide* (London: Academic Press, 2017).

101 D. Levit, J. Ablin, V. Aloush, and I. Yash, "Evaluating Fibromyalgia Symptoms in Transgender Patients," *Arthritis and Rheumatology*, 71 (2019): suppl 10.

102 R. Blanchard, H. E. Barbaree, A. Bogaert et al., "Fraternal Birth Order and Sexual Orientation in Pedophiles," *Archives of Sexual Behavior*, 29 (2000): 463–478.

103 K. Freund and R. J. Watson, "The Proportions of Heterosexual and Homosexual Pedophiles Among Sex Offenders Against Children: An Exploratory Study," *Journal of Sex and Marital Therapy*, 18 (1992): 34–43.

104 R. Blanchard, "Varieties of Autogynephilia and Their Relationship to Gender Dysphoria," *Archives of Sexual Behavior*, 22 (1993): 241–251.

105 E. Dutton and G. Madison, "Gender Dysphoria and Transgender Identity Is Associated with Physiological and Psychological Masculinization: A Theoretical Integration of Findings, Supported by Systematic Reviews," *Sexuality Research and Social Policy*, 18 (2021): 788–799.

to trouble negotiating the social world) and obsessiveness, as a means of comprehending a world to which they are extremely sensitive and thus perceive as disordered and chaotic. This can lead to changes in the sense of self and also to fetishes, seemingly mediated by a high level of environmental sensitivity and ease of sexual arousal, known as "hyper-sexuality."[106] Autism is also an accepted indicator of developmental instability.[107] Autism diagnoses have themselves increased across time and it has been argued that this is at least in part due to genetic factors.[108]

Borderline Personality Disorder (lack of emotional regulation, fear of abandonment and an unstable sense of self) and Narcissistic Personality Disorder also strongly correlate with being transsexual.[109] Indeed, it has been argued that the ferocious reactions of some transwomen to being challenged is an example of "Narcissistic Rage," where the Narcissist must destroy that which threatens to bring down his false self and confront him with his own fragility.[110] There is an extent to which the vulnerable Narcissistic subtype—in contrast to the Grandiose Narcissist, who really feels that he is exalted—is an example of Borderline Personality Disorder.[111]

106 D. Schöttle, P. Briken, O. Tüschen and D. Turner, "Sexuality in autism: hypersexual and paraphilic behavior in women and men with high-functioning autism spectrum disorder," *Dialogues in Clinical Neuroscience*, 19 (2017): 381–393.

107 Woodley of Menie, Sarraf, Pestow and Fernandes, "Social Epistasis Amplifies the Fitness Costs of Deleterious Mutations, Engendering Rapid Fitness Decline Among Modernized Populations," *op. cit.*

108 M. Lynch, "Mutation and human exceptionalism: Our future genetic load," *Genetics*, 202 (2016): 869–875.

109 A. M. Meybodi, A. Hajebi, and A. G. Jolfaei, "The frequency of personality disorders in patients with gender identity disorder," *Medical Journal of the Islamic Republic of Iran*, 28 (2014): 90.

110 A. A. Lawrence, "Shame and Narcissistic Rage in Autogynephilic Transsexualism," *Archives of Sexual Behavior*, 37 (2008): 457–461.

111 S. Euler, D. Stöbi, Sowislo, J. et al., "Grandiose and Vulnerable Narcissism in Borderline Personality Disorder," *Psychopathology*, 51 (2018): 110–121.

Sexual Fantasy and the Self

In transsexuality, there is a sense in which the sufferer is literally sexually attracted to himself, or to the false, perfect female self which he imagines. One transitioning transwoman, who already had breasts, but still had a penis, recalled:

> The first 6 months when my breasts were developing were great. I'd look in the mirror and it wouldn't matter how small the change was, it'd cheer me up. And while I still could, I'd jerk off to my own naked reflection. A standard masturbation technique was to squat in front of the dresser mirror so that my penis was not visible in the reflection. Then I'd concentrate on that while I touched myself. It worked, and I liked the idea of finding my own body to be sexy.[112]

American psychologist Ray Blanchard has proposed that certain fetishes involve an erotic target location error. He distinguishes between "homosexual transsexuality," which emerges at a young age and involves a highly feminised male child feeling that he is female and being attracted to males, and "autogynephilious transsexuality" which usually emerges in early adolescence or later and in which the sufferer is often sexually attracted to females. In the case of "autogynephilous transsexuality," this involves heterosexual males being sexually aroused by the idea of themselves as females and being so obsessed with this fantasy that they take action to try to resemble females and wish to be accepted as females.

Blanchard argues

> that there must be some developmental process that, in most males, keeps such learning 'on track,' perhaps by biasing erotic response towards external rather than internal stimuli, and inherent rather than variable features of the desired object's appearance. When this process fails, the individual may develop an erotic image of himself that includes some, or all, features of

112 A. A. Lawrence, *Men Trapped in Men's Bodies: Narratives of Autogynephilic Transsexualism* (New York: Springer, 2013), 100.

> the desired object [...]

Thus, for every paraphilia, there is a subset of those who are aroused by being the object itself. Certain paedophiles are aroused by the idea of being children. Some people are aroused by amputees, possibly because of the sexual dominance involved in having a relationship with one, so a subset is aroused by the idea of being an amputee. That this "process" has failed would be consistent with developmental instability. These target inversions, however, must be distinguished, argue Czech-Canadian sexologist Kurt Freund (1914–1996) and Ray Blanchard, from, for example, the fantasy of being a child and being "spanked on the buttocks" by his mother's friend, a dominant female, and so, hurt, dominated, placed in a position of sexual submission and humiliated. Such a fantasy reflects, in part, an extreme arousal by being the submissive partner in a relationship with a dominant female: "With masochistic gynaephiles, the same imagery increases the subject's difference from the sexual object (women), in particular, the difference between subject and object in power and control [...]"[113]

Future non-homosexual transwomen begin by masturbating while fantasising that they are women, perhaps wearing women's clothes while masturbating over this fantasy. Thus, they have been referred to as "Men Trapped in Men's Bodies." To quote one psychosexual history, written in the third person:

> By the age of 8, he fantasized about being a girl and

113 K. Freund and R. Blanchard, "Erotic Target Location Errors in Male Gender Dysphorics, Paedophiles, and Fetishists," *op. cit.* It has also been argued that an element of such fantasies, in some cases, may be the eroticization of childhood trauma, to make it more positive, and re-enacting it in order to feel in control of it. See, G. W. Taylor and J. M. Ussher, "Making sense of S&M: A discourse analytic account," *Sexualities*, 4 (2001): 293–314. Autistics are more likely to experience events as trauma: E. Lobregt-van Buuren, M. Hoekert, and B. Sizoo, "Autism, Adverse Events, and Trauma" in A. M. Grabrucker, (Ed.), *Autism Spectrum Disorders* (Brisbane: Exon Publications, 2021). On the evolution of spanking see: E. Dutton and G. Madison, "Spare the Rod and Spoil the Group's Cultural Fitness? Conditions under which Corporal Punishment Leads to Detrimental and Beneficial Outcomes," *Medical Hypotheses,* 145 (2020): 110334.

> wearing girls' clothing nearly every night as he lay in bed. Whenever he did this, his penis became erect. He didn't understand why this happened, and he didn't like it. He found photographs of women's and girls' clothes in magazines and catalogs; he fantasized about wearing these pretty clothes. He fantasized about having long hair and being treated like a girl and having a girl's name; these fantasies also made his penis become erect. He was ashamed of having a penis and wanted it gone. He fantasized about having an operation to remove his penis; ironically, this also caused his penis to become erect. When he was age 9 [...] He tried on panties, bras, slips, and dresses. His penis always became erect, and sticky fluid oozed from its tip.[114]

This person, however, was attracted to females and was relatively masculine. "His favourite subjects in school were geometry and calculus, and his hobby was programming computers."

On many measures, autogynophiles are actually more masculinized, both physically and mentally (with autism being an "extreme male brain") than are controls.[115] In some cases, these do not transition, but they must imagine that they are a woman in order to orgasm during sex with their wives: "I am married to a lovely lady who has been my wife for 23 years. I love her, but the only way I can have sex is to dream I am a woman with breasts and a vagina."[116] Intensity of gender dysphoria decreases after orgasm—after sexual release—strongly implying that is a paraphilia.[117]

114 A. A. Lawrence, *Men Trapped in Men's Bodies: Narratives of Autogynephilic Transsexualism* (New York: Springer, 2013), 4.

115 E. Dutton and G. Madison, "Gender Dysphoria and Transgender Identity Is Associated with Physiological and Psychological Masculinization: A Theoretical Integration of Findings, Supported by Systematic Reviews," *Sexuality Research and Social Policy*, 18 (2021): 788–799.

116 Lawrence, *Men Trapped in Men's Bodies, op cit.*, 116.

117 Lawrence, *Men Trapped in Men's Bodies, op. cit.*, 148.

Increasing LGBT Identification

We would expect that an increase, for partly genetic reasons, in sexual deviance will undermine the optimum adaptive environment, pushing environmentally sensitive people towards deviant forms of sexuality and towards gender dysphoria. Runaway individualism and the influence of "spiteful mutants" would only compound and quicken this process. Consistent with this, there has been a recent significant increase in homosexual identification. In 2015, 2 per cent of young British people claimed to be bisexual. By 2019, it was 16 per cent. Those identifying as "gay or lesbian" increased, over the same period, from 5 per cent to 10 per cent.[118]

These rapid and dramatic changes are clearly due to environmental factors, such as runaway individualism promoting homosexuality and the collapse of institutions that pathologized homosexuality. In the US, as of 2021, 20 per cent of those under the age of 30 identified as "LGBT," but this does not reflect their sexual behaviour: it reflects young women increasingly *identifying* as bisexual or non-binary. Among extreme liberal heterosexuals, 15 per cent identified as LGBT in 2016 but by 2021 it was 33 per cent. In this age-group, liberalism, identifying as LGBT, and depression seemed to be underpinned by a single variable which explained 50 per cent of the variance.[119] This makes sense of evidence that so many traits currently associated with leftism, including being a vegetarian and even dyeing your hair unnatural colours, are associated with depression. In these cases, vegetarianism is a means of virtue-signalling, while dyeing your hair an unnatural colour deals with anxiety by asserting a strong sense of identity, virtue-signals leftist identity via rejection of traditional norms, and signals to the objects of your paranoia that they should stay away, like bright colours on a hoverfly who

118 V. Waldersee, "One in five young people identify as gay, lesbian or bisexual," *YouGov* (3rd July 2019), https://yougov.co.uk/topics/relationships/articles-reports/2019/07/03/one-five-young-people-identify-gay-lesbian-or-bise.

119 E. Kaufmann, *Born This Way? The Rise of LGBT as a Social and Political Identity* (London: CSPI, 2022).

is trying to imitate a (genuinely poisonous) wasp.[120]

Obviously, the normalization of homosexuality, and especially of male homosexuality, creates dysphoria by evoking constant disgust. The sight of two men kissing evokes the same physiological disgust response in the average man as does viewing a bucket of writhing maggots.[121] This response appears to be partly evolved as a pathogen-avoidance mechanism (with homosexual males being high in sexually transmitted diseases),[122] partly because of the strong crossover between the abuse of children and homosexuality which we noted above,[123] and possibly even because of the correlation between homosexuality and mental illness, which, as we have discussed, can be contagious.[124]

As of 2021, over half of American students (52 per cent) taking political majors, such as race or gender studies, identified as LGBT, compared to 25 per cent of students overall and 38 per cent of students at Liberal Arts colleges.[125] One interpretation is that in a highly liberal context being a sexual minority provides you with status in the hierarchy of marginalization, incentivising

120 E. Dutton and E. Kirkegaard, "Blue Hair and the Blues: Dyeing Your Hair Unnatural Colours is Associated with Depression," *Psychreg: Journal of Psychology*, 6:2.

121 B. O'Handley, K. Blair, and R. Hoskin, "What do two men kissing and a bucket of maggots have in common? Heterosexual men's indistinguishable salivary α-amylase responses to photos of two men kissing and disgusting images," *Psychology and Sexuality*, 8 (2017): 177–188.

122 F. van Leeuwen, Y. Ingbar, M. Petersen et al., "Disgust sensitivity relates to attitudes toward gay men and lesbian women across 31 nations," *Group Processes and Intergroup Relations* (2022); B. Everett, "Sexual Orientation Disparities in Sexually Transmitted Infections: Examining the Intersection Between Sexual Identity and Sexual Behavior," *Archives of Sexual Behavior*, 42 (2013): 225–236.

123 G. Gallup, "Attitudes Toward Homosexuals and Evolutionary Theory: The Role of Evidence," *Ethology and Sociobiology*, 17 (1996): 281–284.

124 Homosexuals' levels of mental illnesses such as depression are six times that of heterosexuals of the same gender; they commit suicide at eight times the rate of their heterosexual counterparts; and they are far more likely to engage in drug use and risky sex, something especially true of bisexual males. L. R. Norman, "Bisexuality" in S. Loue, (Ed.), *Mental Health Practitioners' Guide to HIV/ AIDS* (New York: Springer, 2012), 104.

125 Kaufmann, *Born This Way? op. cit.*

younger people to persuade themselves that they are "LGBT." This is more pronounced among females because they are higher in social conformity,[126] likely due to their higher Agreeableness, Conscientiousness and social anxiety when compared to males.[127] Neuroticism has been found to be associated with extrinsic religiousness and thus with social conformity.[128] If the dominant religion is wokeism, they will more strongly conform to it than men. The same is true if the dominant religion is conservative. By the 1990s, older women—raised in a conservative, religious society—were *more* conservative than older men. However, younger women were *less* conservative than males of the same age.[129]

Further, there has been a recent dramatic rise in young people, especially young females, declaring they are "non-binary" or "trans," seemingly as a result of social contagion: something that is fashionable, frequently discussed and, in a sense, status-providing will spread. This is known as "Rapid Onset Gender Dysphoria,"[130] and it seems to rapidly spread through teenage female social groups. One can reasonably assume that earlier generations of such females might have undergone a different mind-body dysphoria, such as anorexia, a condition which has some of the same psychological correlates as gender dysphoria (such as autism) and which correlates with it. As autistics have trouble negotiating the social world, they can develop an unclear sense

126 A. Eagly and C. Chrvala, "Sex Differences in Conformity: Status and Gender Role Interpretations," *Psychology of Women Quarterly,* 10 (1986): 203–220.

127 C. Soto, O. John, S. Gosling, and J. Potter, "Age differences in personality traits from 10 to 65: Big Five domains and facets in a large cross-sectional sample," *Journal of Personality and Social Psychology,* 100 (2011): 330–348.

128 P. Hills, L. Francis, M. Argyle, and C. Jackson, "Primary personality trait correlates of religious practice and orientation," *Personality and Individual Differences,* 36 (2004): 61–73.

129 P. Norris, "Mobilising the 'Women's Vote': The Gender Generation Gap in Voting Behaviour," *Parliamentary Affairs,* 49 (1996): 333–342.

130 L. Littman, "Parent Reports of Adolescents and Young Adults Perceived to Show Signs of a Rapid Onset of Gender Dysphoria," *PLoS ONE,* 13 (2018): e0202330.

of self and can also become highly anxious. Anorexia reflects a delusion about the self, which, nevertheless, "makes sense" of the situation, high anxiety about how one is perceived, and the anxiety-reducing empowerment of control over one's body.[131] A few decades ago, such females might have gone through a phase of believing they were lesbians, perhaps at an all-female school. However, a less individualistic society would have forced them to suppress this.[132]

Suppressing Atheism

Autism tends to correlate with atheism in current samples, for reasons we have already discussed. Autism is strongly elevated among the children of older fathers, due to increasingly low-quality sperm, and thus *de novo* mutation.[133] But, interestingly, the correlation between paternal age and atheism (something that reflects autism, as already noted) only came into existence in American cohorts born in the 1970s. This implies that prior to that, atheistic autistics were successfully suppressing their atheistic beliefs through "effortful control."[134] This implies that society

131 E. Dutton and G. Madison, "Gender Dysphoria and Transgender Identity Is Associated with Physiological and Psychological Masculinization: A Theoretical Integration of Findings, Supported by Systematic Reviews," *Sexuality Research and Social Policy,* 18 (2021): 788–799.

132 Sexuality is considerably more environmental in females than in males. Researchers have theorized that females are evolved to live in harems in which they alloparent each other's children, forming intense and sometimes sexual relationships with other females in the process, with strong parental bonds allowing greater parental stability. For this reason, females have homosexual phases to a greater extent than males. A minority of lesbians, however, are "butches" and are simply highly masculinized. See, A. Jeffery, T. K. Shackelford, V. Zeigler-Hill, J. Vonk and M. McDonald, "The Evolution of Human Female Sexual Orientation," *Evolutionary Psychological Science,* 5 (2019): 71–86.

133 A. Reichenberg, R. Gross, M. Weiser et al., "Advancing Paternal Age and Autism," *Archives of General Psychiatry,* 63 (2006): 1026–1032.

134 M. A. Woodley of Menie, S. Kanazawa, J. Pallesen and M. Sarraf, "Paternal Age is Negatively Associated with Religious Behavior in a Post-60s But Not a Pre-60s US Birth Cohort: Testing a Prediction from the Social Epistasis Amplification Model," *Journal of Religion and Health,* 59

was pushing people in an adaptive direction—we have noted that religiousness is adaptive, at the individual and group level—and that it gradually stopped doing so around 1970.[135]

What would remain would be assorted groups who, for genetic reasons, were strongly adaptive and or otherwise strongly resistant to such inclinations and, as we have discussed, it is these people who disproportionately breed. They would do so because they are conservative and religious for strongly genetic reasons or because, being relative r-strategists and with low IQ, they are insensitive to the increasingly maladaptive environment, again likely for significantly genetic reasons. The former category has strong genetic mental and physical health. The latter category has the opposite, and it is to dysgenic physical health that we will now turn.

(2020): 2733–2752.

135 Where high functioning autistic people are religious, we might expect them to be attracted to what appear, at least, to be extremely rational forms of religiosity, where, for example, the existence of God seems to be logically proven. This might reflect low IQ, which correlates with religiosity, or high Conscientiousness outweighing the low empathy, autism traits. It might also reflect high anxiety, which is associated with religious experiences and periods of religious fervour, overwhelming these traits. See, Dutton and Van der Linden, "Why is Intelligence Negatively Associated with Religiousness?" *op. cit.* and E. Dutton, *Religion and Intelligence: An Evolutionary Analysis* (London: Ulster Institute for Social Research, 2014). We might also expect them to be attracted to highly ritualistic forms of religiosity, to the extent that these provide them with structure, order and predictability. In this regard, it has been suggested that some leading monks and nuns may have been autistic. See, for example, P. Ranft, "Ruminations on Hildegard of Bingen (1098–1179) and autism," *Journal of Medical Biography*, 22 (2014): 107–115. In this sense, autistic traits could stay in populations as a side effect of group-selection for an optimum number of geniuses (as explored elsewhere in this book), but also as a side effect of an optimum number of religious renouncers who inspire group-oriented religiosity.

Seven

The Rise and Fall of Genetic Physical Health

A Monstrous Birth

In about 1512, in Ravenna in Italy, a woman, apparently a nun who had been impregnated by a wayward Friar, gave birth to a "monster." Accounts of it were doubtless embellished, but it supposedly had horns, cloven feet, and an eye in the middle of one of its legs, which were fused together. It was also a hermaphrodite. Modern scientists have suggested various tentative diagnoses for the severe congenital disorder from which this doomed baby suffered. Two years later in Bologna, a little girl who lived long enough to be baptised "Maria," was born with two faces, three eyes, and a woman's vulva on her forehead. She died a few days later.[1] These were both extreme examples of what were known as "monstrous births." They fascinated people, as they were, of course, horrifying and very unusual. Such "monsters" would be high in mutational load, and they would, therefore, have very poor immune systems and assorted comorbidities, meaning that they would die relatively young and only the most pronounced cases, such as the "Monster of Ravenna," would become widely known.

As discussed earlier, since the Industrial Revolution, child mortality has collapsed from about 50 per cent to less than 1 per cent, and this has co-occurred with an increasingly easier climate.

1 See, A. W. Bates, *Emblematic Monsters: Unnatural Conceptions and Deformed Births in Early Modern Europe* (Amsterdam: Rodopi, 2005).

As a result, purifying selection against mutation has collapsed. Those with only a moderately high level of mutational load pass on their genes and, indeed, those with a genetic propensity to obesity, strokes and heart attacks outbreed healthy controls, due to the ability of medical science to defer death from these issues to after the age at which people can have children.[2] Not only that, but intelligence is negatively associated with genes for heart disease and high blood pressure.[3] People with severe congenital problems may not breed themselves, but they can certainly be a significant presence in society, creating a strong evolutionary mismatch, in part due to the disgust-response which they evoke, this likely being an adaptation to avoiding disease.[4] The impact of such people can range from the noticeable presence of very severely disabled people, who require full time care throughout their lives and who create dysphoria via moderate disgust, to those who would have been regarded as manifest "monstrous births" in the sixteenth century and who evoke extreme disgust.

This process, of the handicapped being increasingly present in the community, began in Britain in the 1980s with a policy known as "Care in the Community" whereby the physically and mentally handicapped were increasingly removed from institutions, which segregated them from society, and looked after, as much as possible, in their own homes.[5] The same process occurred over the same period in the US and in many other developed countries, with little consideration for what has been termed "the aesthetics of disability"—the unnerving behaviour of severe autistics, for example—and its psychological impact on

2 D. Hugh-Jones and A. Abdellaoui, *Natural Selection in Contemporary Humans is Linked to Income and Substitution Effects* (University of East Anglia, 2021).

3 I. Deary, S. Harris and W. D. Hill, "What genome-wide association studies reveal about the association between intelligence and physical health, illness, and mortality," *Current Opinion in Psychology*, 27 (2019): 6–12.

4 M. Oaten, R. Stevenson, and T. Case, "Disease avoidance as a functional basis for stigmatization," *Philosophical Transactions of the Royal Society of London: B*, 366 (2011): 3433–3452.

5 A. Llewellyn, L. Agu and D. Mercer, *Sociology for Social Workers* (Cambridge: Polity Press, 2008), 268.

the general population.[6] Their segregation, to the extent that they even survived childhood at the time, began in earnest with the Industrial Revolution, which resulted in many women working outside the home, overcrowding, the weakening of community ties, and work being faster and more disciplined than was farm labour. In such a context, even the mildly physically and mentally handicapped were a serious burden both on their families and on the rate payer.[7] Prior to this time, the "village idiot" would engage in farm labour along with the rest of the largely illiterate rural community.[8]

It may be that one of the conditions that afflicted the "Monster of Ravenna" was "sirenomilia," otherwise known as "Mermaid Syndrome." In this condition, the legs are fused, meaning the sufferer lacks any means to pass out waste. They often lack vital organs such as a uterus, a bladder, a small intestine, and other crucial organs, such as the kidneys, don't function properly. As with the "Monster of Ravenna," they usually die shortly after birth. Shiloh Pepin (1999–2009) of Kennebunkport, Maine, an extraordinarily eloquent and friendly girl who suffered from Mermaid Syndrome, amazed doctors by living to the age of ten before finally succumbing to pneumonia.[9]

The Rise of Developmental Instability

Shiloh Pepin is obviously an extreme example. She is one of only three people with Mermaid Syndrome who have ever made it beyond the first days of life. But weakened selection pressure, and the consequent rising mutational load, has had many far more subtle effects, as we saw with regard to the case of Nicky Ratcliffe earlier. Nicky was pretty; but there is evidence that our

6 J. E. Harris, "The Aesthetics of Disability," *Columbia Law Review*, 119 (2019): 895–972.

7 S. Carnaby, *Designs for Living: A Comparative Approach to Normalisation for the New Millennium* (London: Routledge, 2018).

8 D. Race, "The Historical Context," in D. Race, (Ed.), *Learning Ability: A Social Approach* (London: Routledge, 2012), 25.

9 B. Brown and L. Sher, "Girl Born With Rare 'Mermaid' Condition Dies," *ABC News* (26th October 2009), https://abcnews.go.com/2020/shiloh-pepin-girl-born-rare-mermaid-condition-dies/story?id=8920709.

rising mutational load means that even this sign of relative genetic health is becoming rarer. A face will be attractive, in part, because it is symmetrical. As discussed, we are evolved to find symmetry attractive as it betokens healthy genes; the ability to maintain a symmetrical phenotype in the face of disease and a lack of inherited mutations. However, it has been found from a representative sample of skulls that the average faces of Westerners have become less symmetrical, in other words less attractive, since the nineteenth century. This change is also consistent with falling genetic intelligence over this period. There is a weak correlation between facial symmetry and intelligence, due to both having previously been under positive selection and due to both being a reflection of low mutational load.[10] Increasing facial asymmetry implies both increased mutational load directly, as reflected in the face, but also a more general "developmental instability." Our immune systems are becoming weaker on average due to rising mutational load, so we must use proportionately more of our bio-energetic resources to fight off disease, making us less able to produce a symmetrical phenotype. That said, we are exposed to far fewer microbes than our ancestors. This means that less bio-energetic resources are needed. Thus, the skulls are also likely to simply reflect rising mutational load, as expressed in the face. The face, in this sense, is an "honesty signal" of genetic quality, as with the peacock's tail which we discussed earlier.

We observed in the last chapter that homosexuality and paedophilia can be regarded as examples of developmental instability due to the way in which they correlate with poor genetic health. Another example of developmental instability is left-handedness. It correlates with poor physical and mental health and, weakly, with low intelligence. "Sinistrality," as it is also known, implies a strongly asymmetrical brain, or atypical development of fine motor skills in the right hemisphere, as right-handedness is the norm and, in general, only this kind of asymmetry would produce left-handedness. It thus implies either relatively high

10 M. A. Woodley of Menie and H. Fernandes, "The secular decline in general intelligence from decreasing developmental stability: Theoretical and empirical considerations," *Personality and Individual Differences,* 92: 194–199.

mutational load or a sub-optimal environment at a very early stage of development. Analyses of photographs of people waving, and thus using their dominant hand, with modern surveys indicate that left-handedness has been increasing across time.[11] This would also be congruous with weakened selection pressure leading to the kind of psychological changes that we explored in the last chapter, including declining intelligence and also declining group selection, specifically with regard to national-level group-selection. The result of the latter is weaker selection for highly cooperative people, who will tend to be lower in aggression and testosterone. Consistent with this, left-handedness is higher among men, and it is also higher among aggressive, warlike tribes who are subject to only weak selection pressures. Sinistrality is about 12 per cent in Europe, compared to 20 per cent among the violent Yanomamö of Venezuela.[12] These groups, in the context of warfare, would be less internally cooperative and less able to incrementally innovate, due to lack of impulse control and lack of intelligence and future orientation. As such, they would be less able to present a successful and united front in the face of the enemy. Group-selection would also be lower in these ecologies due to relative resource abundance.

In essence, Western populations are becoming sicker over time for genetic reasons. Poor mental health is part of this and even low religiosity, with the traditionally religious having high physical and mental health for partly genetic reasons.[13] Certainly, religious people—in the sense of those who collectively worship a moral god—are less vulnerable to depression, apparently in part for genetic reasons.[14] They also have superior immune

11 M. A. Woodley of Menie, H. Fernandes, S. Kanazawa, and E. Dutton, "Sinistrality is associated with (slightly) lower general intelligence: A data synthesis and consideration of secular trend data in handedness," *HOMO: Journal of Comparative Human Biology*, 69 (2018): 118–126.

12 E. Dutton, G. Madison, and D. Van der Linden, "Why do high IQ societies differ in intellectual achievement? The role of schizophrenia and left-handedness in per capita scientific publications and Nobel prizes," *Journal of Creative Behavior*, 54 (2020): 871–883.

13 H. Koenig, "Religion, Spirituality, and Health: The Research and Clinical Implications," *ISRN Psychiatry* (2012): 28730.

14 M. Anderson, L. Miller, P. Wickramaratne et al., "Genetic Correlates

systems, even when controlling for many environmental factors that might impact this.[15] But let us turn to the evidence of crumbling genetic health.

The Rise of Cancer and Other Partly Genetic Disorders

If the human genome were crumbling, as is being proposed, then we would expect to see a rise across time in people with significantly genetic disorders. Clearly, many health conditions—such as obesity and heart disease—have a significant environmental component, though some people will be more genetically prone to them than others. Heart disease, for example, has a heritability of around 0.5,[16] while the heritability of obesity is around 0.55.[17]

One issue which is consistent with a rise in mutational load is the rise in prevalence of cancer. As discussed earlier, in Mid-Victorian England very few people ever suffered from cancer, even controlling for age. A medical student at Charing Cross Hospital who examined a cadaver that had died of lung cancer in 1869 was told by his lecturer that lung cancer was: "one of the rarer forms of a rare disease. You may probably pass the rest of your students' life without seeing another example of it." There is also evidence that breast cancer, to the extent that it existed in Victorian England, was far less aggressive and fatal than is the case today. In addition, heart disease did not begin to appear as a prominent cause of death until 1857, strokes were generally not fatal, and both were regarded as diseases of extreme old age. Certainly, a component of this change is lifestyle: we do less exercise than the Mid-Victorians, alcoholic drinks are stronger, our diet is lower in fruit and vegetables, and it is higher in sugar (a

of Spirituality/Religion and Depression: A Study in Offspring and Grandchildren at High and Low Familial Risk for Depression," *Spirituality and Clinical Practice*, 4 (2017): 43–63.

15 H. Koenig, J. Hays, L. George et al., "Modelling the cross-sectional relationships between religion, physical health, social support, and depressive symptoms," *American Journal of Geriatric Psychiatry*, 5 (1997):131–45

16 R. McPherson and A. Tybjaerg-Hanson, "Genetics of Coronary Artery Disease," *Circulation Research*, 118 (2016): 564–578.

17 B. Herrera and C. Lindgren, "The Genetics of Obesity," *Current Diabetes Reports*, 10 (2010): 498–505.

luxury in Mid-Victorian England), meat, and fat. It is specifically lower in offal, offal being healthier than the tastier cuts of meat which now comprise more of our diet.[18] However, it is surely also relevant that in the absence of purifying selection, mutational load builds up each generation and the child mortality rate in England in 1850 was 16 per cent compared to less than 1 per cent today.[19]

The rise in cancer prevalence is thus very informative in terms of mutation accumulation. It is a significantly heritable condition that is often polygenic in origin; it involves many genes failing to replicate properly in a particular part of the body. Specifically, cancer is caused by "somatic mutations" (cell alterations post-conception due to problems in cell replication) rather than "germ-line mutations," these being mutations inherited directly from the parents. These somatic mutations build up slowly as we get older (though genes will influence how quickly, how severely and where on the genome), and cancer arises when a single cell acquires a bad enough combination of somatic mutations. There are some cases of inherited cancer susceptibility syndromes where one copy of a tumour-suppressor gene is defective in all cells of the body, and cancer arises easily when the second, intact, copy of that tumour-suppressor gene gets knocked out by a somatic mutation. This means that some cells grow abnormally—they become cancerous—and they can spread to other parts of the body, causing vital organs to cease functioning properly. In addition, certain kinds of cancer are strongly genetic, such as certain kinds of breast and ovarian cancer.[20]

Using the prevalence of cancer as a marker of rising mutational load does, it should be said, have various limitations. We have an aging population, and the body degenerates with age render-

18 P. Clayton and J. Rowbotham, "How the Mid-Victorians Worked, Ate and Died," *International Journal of Environmental Research and Public Health*, 6 (2009): 1235–1253.

19 E. J. T. Collins, *The Agrarian History of England and Wales, 1850–1914, Volume VII, Part 2* (Cambridge: Cambridge University Press, 2000), 1268.

20 S. Rowley, L. Mascarenhas, L. Devereux et al., "Population-based genetic testing of asymptomatic women for breast and ovarian cancer susceptibility," *Genetics in Medicine*, 21 (2019): 913–922.

ing cancer more likely, and we are becoming better at diagnosing disorders. So any relevant studies have to control for or adjust for these factors. When we do this, we find that worldwide, based on World Health Organization Data from 173 countries, the earlier a country industrialized the more prevalent cancer is. This is true in people aged 0 to 49 and in people of all ages. Crucially, to those who would insist that the relationship is due to environmental factors: "This relationship remains significant when GDP, life expectancy of older people (e50), obesity, physical inactivity, smoking and urbanization are kept statistically constant for fifteen (15) out of twenty-seven (27) individual cancers incidence rates." These strongly genetic forms of cancer are 5.7 times more common in the most industrialized nations than in the least industrialized nations.[21] So, there is little room for doubt: cancer has become more prevalent since industrialization due to weakening purifying selection, something that has resulted in rising mutational load.

It is not just with cancer that we find this. Celiac disease is a condition whereby people have an adverse reaction to gluten. In other words, it is a kind of allergy or intolerance. A study comparing a cohort from 1978–1980 and from 2000–2001 found a significant increase in the prevalence of the condition and further found that it could not be put down to better diagnosis.[22] The condition is also "strongly heritable," so the rise is likely to involve some kind of genetic component.[23] As discussed above, we know that obesity and heart disease must be increasing for at least partly genetic reasons because they are positively correlated with fertility. Indeed, a general index of genetic poor health—based on the UK Bio-bank of 409,629 subjects—is pos-

21 W. You and M. Henneberg, "Cancer incidence increasing globally: The role of relaxed natural selection," *Evolutionary Applications,* 11 (2017): 140–152.

22 S. Lohi, K. Mustalahti, K. Kaukinen et al., "Increasing prevalence of coeliac disease over time," *Ailmentary Pharmacology and Therapeutics,* 26 (2007). 1217–1225.

23 A. L. King and P. J. Ciclitira, "Celiac disease: strongly heritable, oligogenic, but genetically complex," *Molecular Genetics and Metabolism,* 71 (2000): 70–75.

itively correlated with fertility.[24] The increase in obesity has been stronger the more industrialized a country is. Some of this effect remains even when controlling for the kind of environmental variables that are associated with obesity.[25] This would mean that under harsh conditions, an optimally adapted human would put on weight in times of plenty to survive times of famine, but he would also be evolved to lose excessive weight in order not to waste energy and damage the body by carrying excess weight. Weakened selection has caused more and more people to deviate from this optimum. Furthermore, Type 1 Diabetes, which is strongly a matter of genetics rather than of lifestyle, is increasing globally.[26]

Under harsh conditions, myopia—short-sightedness—is clearly disadvantageous, and those who suffer from it are more likely to make fatal mistakes. Over the last 25 years, analysis of the UK Biobank indicates a continuous increase in alleles that are associated with myopia. In the UK alone this is sufficient to add more than 100,000 new cases of myopia per generation. This is slightly complicated by the fact that myopia itself appears to be negatively associated with fertility, possibly because there is a genetic correlation between myopia and intelligence,[27] and intelligence is negatively associated with fertility. But, even so, the polygenic nature of myopia and the failure to select out the assorted mutant genes that cause it mean that it is increasing

24 Hugh-Jones and Abdellaoui, *Natural Selection in Contemporary Humans is Linked to Income and Substitution Effects, op. cit.*

25 W. You and M. Henneberg, "Relaxed natural selection contributes to global obesity increase more in males than in females due to more environmental modifications in female body mass," *PLOS ONE*, 13 (2018): e0199594.

26 W. You and M. Henneberg, "Type 1 diabetes prevalence increasing globally and regionally: The role of natural selection and life expectancy at birth," *BMJ Open Diabetes Research and Care*, 4 (2016): e000161.

27 D. Czedpita, E. Lodygowska, and M. Czedpita, "Are children with myopia more intelligent? A literature review," *Annales Acadamaie Medicae Stetinensis*, 54 (2008): 13–16; K. Williams, P. Hysi, E. Yonova-Doing, O. Mahroo et al., "Phenotypic and genotypic correlation between myopia and intelligence," *Scientific Reports*, 7 (2017): 45977.

for genetic reasons.[28] Due to the close relationship between the eyes and the brain, a single genetic mechanism may impact both brain size (with brain size weakly correlated with intelligence) and eye-size, resulting in the genetic relationship between myopia and intelligence.[29]

All of this is congruous with a comparison between modern and ancient genomes (using samples taken from ancient skeletons of a set age). It appears that the genomic health of humans improved over time, indicating that they were, in general, continuously subject to harsh purifying selection. Thus, the genetic health of the Altai Neanderthal is worse than that of 97 per cent of modern humans. But then there is a sudden change. Ancient people who lived within the last few thousand years display on average healthier genomes than do modern day humans.[30] As Danish researcher Emil Kirkegaard, who has previously explored these studies, has argued, this would be congruous with humans—although there would be humps and bumps due to climate change and other factors—having been under intense purifying selection until the Industrial Revolution, at which point selection pressures collapsed.[31]

The Rise of Infertility

From an evolutionary perspective, although one can pass on one's genes indirectly through kin selection and group selection, the key aim of an organism is to directly pass on its genes. As my secondary school biology teacher once told us, "The purpose of life is to breed. There's nothing else." Genetically unfit organisms are, under harsh Darwinian conditions, prevented from so do-

28 E. Long and J. Zhang, "Natural selection contributes to the myopia epidemic," *National Science Review,* 8 (2021): https://doi.org/10.1093/nsr/nwaa175.

29 E. Miller, "On the correlation of myopia and intelligence," *Genetic, Social, and General Psychology Monographs,* 118 (1992): 361–383.

30 A. Berens, Taylor Cooper, and J. Lachance, "The Genomic Health of Ancient Hominins," *Human Biology,* 89 (2017): 7–19.

31 E. Kirkegaard, "Heath Dysgenics: A Very Brief Review," *Clear Language, Clear Mind* (9th August 2017), https://emilkirkegaard.dk/en/2017/08/health-dysgenics-a-very-brief-review/.

ing via a combination of natural selection and sexual selection. But, clearly, putting aside infections and sexually transmitted diseases, being so unhealthy that you cannot fertilize a female or become pregnant under natural conditions would betoken high mutational load. Thus, we would expect it to become a growing problem as deleterious mutations accumulate over time. As mortality has collapsed before and during prime reproductive years, fertility should be a good marker of fitness.[32]

Congruous with this, there is evidence that infertility has a genetic component. Western males have undergone a dramatic fall in testosterone levels since the 1970s, which would be expected to negatively impact sperm production.[33] Consistent with this, sperm counts have fallen enormously since the 1970s.[34] Males have also become physically weaker, as measured by grip strength.[35] There is no sound evidence linking the falling sperm count to chemical influences. Indeed, exposure to toxins has been decreasing over time.[36] It is crucial to emphasize that sperm quality is an accepted marker of general health. So, if sperm quality is falling, then general health is doing the same, though this may be mediated by obesity interfering with testosterone production in males.[37]

It is difficult to unpick the data on rising infertility, as part

32 Y. Kim and J. Lee, "The genetics of human fertility," *Current Opinion in Psychology*, 27 (2019): 41–45.

33 A. M. Andersson, T. K. Jensen, A. Juul, A. et al., "Secular decline in male testosterone and sex hormone binding globulin serum levels in Danish population surveys," *Journal of Clinical Endocrinology and Metabolism*, 92 (2007): 4696–4705.

34 H. Levine, N. Jørgensen, N., A. Martino-Andrade et al., "Temporal trends in sperm count: A systematic review and meta-regression analysis," *Human Reproduction Update*, 23 (2017): 646–659.

35 E. Fain and C. Weatherford, "Comparative study of millennials' (age 20–34 years) grip and lateral pinch with the norms," *Journal of Hand Therapy*, 29 (2016): 483–488.

36 M. Sarraf, M. A. Woodley of Menie, and C. Feltham, *Modernity and Cultural Decline: A Biobehavioural Perspective* (London: Palgrave Macmillan, 2019), 198, FN 3.

37 Sarraf, Woodley of Menie and Feltham, *Modernity and Cultural Decline, op. cit.*, 198, FN 2.

of this reflects people heavily delaying their fertility for cultural reasons. This makes it difficult to compare rates of infertility across time. In addition, it may be that humans, like other pack animals, go through hormonal cycles across many decades, as an adaptation to be less predictable to predators and prey.[38] But the key cause of male infertility is poor sperm quality or low sperm count, and sperm quality (as measured by sperm count) is certainly becoming poorer across time, consistent with rising mutational load, though there may be other explanations. Clearly, males postponing having children to older ages would make this problem of rising mutation even worse, for reasons already outlined.

The Survival of the Least Fit

So, it is clear that humans in industrialized societies are becoming less genetically healthy due to the gradual mutation accumulation permitted by the extreme weakening of purifying selection. And—as we have already explored—physical fitness correlates with mental fitness and all other kinds of fitness, meaning there is a general fitness factor, and we are becoming less fit.

Our ability to successfully make medical interventions, which has increased hugely since the nineteenth century, clearly aids this process. If a mother cannot give birth normally then both she and the offspring are likely to die, limiting the fertility of the mother and potentially destroying that of the offspring. Increasingly successful Caesarian sections allow the mother, who likely has elevated mutational load, to have more offspring, to pass on her mutational load and to pass on the inability to naturally give birth to any girls she may have. A mother born by Caesarian is 55 per cent more likely than controls to require a Caesarian herself and the need for Caesarian births has increased across time.[39]

As discussed earlier, this situation not only means that those who have moderate mutational load can pass on their genes,

38 This is argued in J. Miettinen, "Cyclical Metapopulation Mechanism Hypothesis." Unpublished Manuscript.

39 M. Tollånes, S. Rasmussen, and L. Irgens, "Caesarean section among relatives," *International Journal of Epidemiology*, 37 (2008):1341–1348.

contributing to rising mutational load, but also that those with serious genetic disorders that would generally be associated with infertility become an increasing presence in society. It also means that those who develop serious disorders as they age have a much higher life expectancy, resulting in increasing morbidity and growing numbers of people requiring full time care. The societal effects of increasing genetic ill-health will, in turn, be amplified by improved medical care, which allows affected individuals to survive much longer. The genetic effect is very slow—cumulative over many generations of relaxed selection. However, the effect of improved medical care is almost instantaneous, as in the example of Down's Syndrome. This difference between the rapidity and magnitude of environmental and genetic effects tends to obscure the importance of genetics, because most people are ill-prepared to consider consequences that become real but only many generations hence. We saw earlier that in 1945, the life expectancy of somebody born with Down's Syndrome was about 12 years. Since the development of antibiotics and other medical innovations, it has increased to about 60 years.

The life expectancy of those with other chronic and infertility-causing disorders is also rising. Before the 1980s, more than half of people with cystic fibrosis (an autosomal recessive disease affecting about 1 in 2,500 new-borns) did not reach their 20s. By the early 2000s, around half died before the age of 37. Their median life expectancy has now increased to 44. As a result, it has been predicted that in the coming years there will be a 75 per cent increase in the prevalence of cystic fibrosis sufferers in the population, all of whom will require frequent medical care.[40] The same could be said of sufferers of a large variety of other conditions in which patients are unlikely to have children but will be a growing presence in society. Another example is those that suffer from Prader-Willi Syndrome, a rare condition caused by loss of specific genes from Chromosome 15. In 70 per cent of cases, this faulty chromosome is inherited from the father, something

40 P.-R. Burgel, G. Bellis, H. Olesen et al., "The Provision of Care for Adults with Cystic Fibrosis in Europe," *European Respiratory Journal*, 46 (2015): 133–141.

known as "paternal deletion."[41] Prader-Willi Syndrome is marked by mild mental retardation, behavioural problems, poor muscle tone, facial dysmorphia reminiscent of mild Down's Syndrome, short stature, infertility and, most significantly, a constant desire to eat, such that the appetite is never sated.[42] As a result, people with this condition tend to eat themselves to death. About 80 per cent of known sufferers have lived to the age of 40.[43] The British model Katie Price (b. 1978) has a son called Harvey who has Prader-Willi Syndrome.[44]

Pre-Natal Screening and "Wrongful Life"

Pre-natal screening was developed in the 1970s.[45] The process has become increasingly reliable since then and abortion became legal in the UK, for example, in 1967. The abortion rate for foetuses with Down's Syndrome has gradually increased. In Britain, by 2010, 90 per cent of foetuses with the condition were being aborted.[46] Thus, there was a time, and I remember this as a child, when it was not uncommon to see adults with Down's Syndrome. They are now a much rarer sight because such a high percentage of foetuses with the condition are terminated. I have noticed that they are more common in Finland, consistent with the Finnish abortion rate for Down's Syndrome being just 50 per cent in 2008.[47] Of course, most mutation occurs on the father's

41 A. Wickens, *Introduction to Biopsychology* (London: Pearson Education, 2009), 187.

42 S. Cassidy, S. Schwartz, J. Miller, and D. Driscoll, "Prader-Willi syndrome," *Genetics in Medicine,* 14 (2012): 10–26.

43 M. Butler, A. Manzardo, J. Heinnemann et al., "Causes of death in Prader-Willi syndrome: Prader-Willi Syndrome Association (USA) 40-year mortality survey," *Genetics in Medicine,* 19 (2017): 635–642.

44 K. Price, *Love, Lipstick and Lies* (London: Random House, 2013), 125.

45 I. Löwy, "Prenatal diagnosis: the irresistible rise of the 'visible fetus,'" *Study in History and Philosophy of Biological and Biomedical Sciences,* 47 (2014): 290–299.

46 *TheJournal.ie, FactCheck:* "Are 90% of babies with Down syndrome in Britain aborted?" (4th February 2018), https://www.thejournal.ie/fact-check-babies-abortion-3823611-Feb2018/.

47 *YLE,* "Down syndrome terminations rise in Finland," (28th October

sperm, so increasing paternal age in itself increases the likelihood of genetic poor health in future generations, though rising maternal age makes a lesser contribution.

However, pre-natal screening has resulted in eugenic abortion practices among some ethnic groups for some conditions. In Cyprus, in 1980, pre-marital screening for thalassemia, a relatively widespread genetic disorder in Cyprus, became mandatory such that couples were counselled about whether their offspring might develop the condition. In 1984, pre-natal diagnosis commenced and babies with the condition began to be routinely aborted, resulting in there now being very few sufferers of thalassemia—a blood disorder where insufficient haemoglobin is made—in Cyprus.[48] Pre-natal screening is common in Israel not only for thalassemia but also for Tay-Sachs Disease, which has a relatively high prevalence among Ashkenazi Jews. The child is born normal but gradually the condition stops the nerves from working properly, leading to retardation, seizures, blindness, deafness and, eventually, death, usually by the age of about five. Screening—and consequent abortion—has resulted in a significant reduction in the prevalence of these conditions in Israel.[49] Where both couples are carriers, they are counselled not to marry each other. Many eugenic practices are particularly widely accepted in Israel, having been espoused by prominent early Zionists such as Hungarian Jewish physician Max Nordau (1849–1923). According to Israeli sociologist Meira Weiss: "In Israel, the Zionists' eugenics turned into a selective pre-natal policy backed by state-of-the-art genetic technology."[50]

Courts in most countries where such an idea has been argued

2018), https://yle.fi/uutiset/osasto/news/down_syndrome_terminations_rise_in_finland/10479878.

48 G. Bozkurt, "Results from the North Cyprus thalassemia prevention program," *Hemoglobin*, 31 (2007): 257–264.

49 J. Zlotogora, R. Carmi, B. Lev, and S. Shalev, "A targeted population carrier screening program for severe and frequent genetic diseases in Israel," *European Journal of Human Genetics*, 17 (2009): 591–597.

50 Quoted in: J. Glad, *The Future of Human Evolution: Genetics in the Twenty-First Century* (Schuylkill Haven, PA: Hermitage Publishers, 2008), 69.

for by lawyers have rejected the concept of "Wrongful Life", the idea that seriously mentally and physically disabled people have a "right to non-existence," due to the intense suffering their lives entail. It is averred by the courts that life is uniquely precious and, anyway, such a right would violate the interests of disabled people, implying that their lives were not of value. In Germany, the courts have accepted the notion of "Wrongful Birth"—that the mother was misinformed, did not have an abortion, had a severely disabled child and deserves substantial compensation to look after the child—but not "Wrongful Life."

The Israeli courts, by contrast, ruled in 1986 that there was such a concept as "Wrongful Life." An Israeli woman had undergone genetic counselling because she was concerned about having sons with Hunter's Disease, an X-linked condition where the body cannot properly breakdown sugar molecules, affecting all organs and tissues. The result, though there are also milder forms, is progressive severe mental retardation, mobility difficulties and progressive deafness.[51] The disease existed in her family, but she was wrongly told by a genetic counsellor that the condition could not be inherited by her children. She duly had a son with Hunter's Disease and sued the counsellor. The judges ruled that it was better not to exist than to live with a severe impairment, so the mother won the case.[52]

The Cost of Genetic Diseases

These rare diseases are all associated with both a direct and indirect substantial economic burden on society.[53] The burden of caring for the victims of genetic disorders is an important social consequence of genetic ill-health that needs to be considered in addition to the suffering of the affected individuals themselves.

51 S. Natelson, *Principles of Applied Clinical Chemistry: Chemical Background and Medical Applications. Volume 3: Plasma Proteins in Nutrition and Transport* (New York: Springer, 2012), 138–139.

52 Y. Hashiloni-Dolev, *A Life (Un)Worthy of Living: Reproductive Genetics in Israel and Germany* (New York: Springer, 2007), 122–123.

53 A. Angelis, D. Tordrup, and P. Kanavos, "Socio-economic burden of rare diseases: A systematic review of cost of illness evidence," *Health Policy,* 119 (2015): 964–979.

This is money that could be spent, for example, on nurturing genius and, thus, slowing down civilizational decline, making the process more bearable, a possibility to which we will return later.

A study in 2014 found that 42 per cent of healthcare insurance claims in the US made on behalf of children were for patients with "single-gene diseases or chromosomal abnormalities."[54] In 2016, 10 per cent of the budget of Britain's National Health Service was spent on services for those with rare genetic diseases, yet such people were only 6 per cent of the British population.[55] In 1987, the lifetime cost to the NHS of caring for one patient with Down's Syndrome was double that of examining 100 amniotic fluids and terminating one foetus.[56] The average cost, in 2007, to the British taxpayer of a severely disabled person, who has to live in an institution rather than with their family, was £696 per week. It was £171 per week for those who live with their families but whose families claim government support.[57] Average weekly earnings in the UK in 2007 were £513 per week.[58]

As of 2017, the British government was spending £50 billion annually supporting disabled people and those with serious health conditions.[59] This was about 6 per cent of the entire na-

54 K. Miller, R. Hoyt, S. Rust et al., "The Financial Impact of Genetic Diseases in a Pediatric Accountable Care Organization," *Frontiers in Public Health*, 8 (2020): 58.

55 A. Simpson, *The Hidden Costs of Rare Diseases: A Feasibility Study* (London: Genetic Alliance UK, 2016), 6.

56 D. Shickle, "The Wilson and Junger Principles of Screening and Genetic Testing" in R. Chadwick, D. Shickle, H. ten Have and U. Wiesing, (Eds.), *The Ethics of Genetic Screening* (New York: Springer, 1999), 21.

57 D. Baron, I. Molosankwe, R. Romeo, and A. Hassiotis, "Urban adolescents with intellectual disability and challenging behaviour: costs and characteristics during transition to adult services," *Health and Social Care in the Community*, 21 (2013): 283–292.

58 H. Wise, "Wages should finally surpass 2007 peak this month but 30-somethings still carry 'wage scars' of the recession," *This Is Money* (8th November 2019), https://www.thisismoney.co.uk/money/news/article-7663931/Average-wages-finally-surpass-2007-peak-month.html.

59 T. Miles, "Britain 'going backwards' on rights of disabled, says U.N. committee," *Reuters* (31st August 2017), https://www.reuters.com/article/us-britain-disabled-idUSKCN1BB2AO.

tional budget for that year, and almost the entire budget deficit for that year.[60] As of 2021, 8.7 per cent of the population were claiming "disability benefit," meaning that they were of working age but were apparently unable to work. In poor boroughs, such as Blackpool, this was around 15 per cent. In wealthy boroughs, such as Guildford, it was less than 5 per cent.[61]

Until relatively modern times, however, societies, despite being under high levels of purifying Darwinian selection, have acted to deliberately purge those with poor genetic health from their number even when they do manage to survive. It is to this process of "eugenics" which we will now turn.

60 H. M. Treasury, *Spring Budget, 2017* (London: HMSO, 2017).

61 E. Kirk-Wade, *UK Disability Statistics: Prevalence and Life Experiences* (London: House of Commons, 2022).

Eight

The History and Rise of Eugenics

The Runt of the Litter

In October 1989, my family bought a dog; an English Springer Spaniel whom we called "Chipper." He was a fantastically good-natured and friendly dog and his premature death, aged nearly 8 from epilepsy, plunged us into prolonged grief. However, it seems that Chipper's mother, in her instinctive canine way, *knew* that this was exactly what was going to happen. Chipper was the "runt of the litter." He was rejected by his mother, such that, as a newly born puppy, he was raised entirely by humans and was bottle-fed.

The result was that, when we first got him and started taking him for walks, he had no idea how to behave around other dogs. Chipper would just bound up to them and lick them, this eliciting a positive response from humans. Many of these dogs, being pack animals with notions of status and deference, regarded this as an act of "disrespect" by a low status dog—he was, after all, only a puppy—and made their displeasure perfectly clear. Gradually, Chipper learned how to behave around other dogs. He also developed epilepsy and by 1997 this had become so severe that the vet put him on life support and told us that each fit would trigger another fit until he died. He was, therefore, put down.[1]

1 Epilepsy is known to be elevated among English Springer Spaniels, compared to other breeds of dog, and it has a genetic basis. L. De Rusio and K. Munana, *A Practical Guide to Seizure Disorders in Dogs and Cats* (Palm

Heartless as it was, Chipper's mother's instinctive rejection of him as a new-born was evolutionarily adaptive behaviour. She correctly intuited that there was something genetically wrong with him and that it would be a waste of her bio-energetic resources, which she could invest entirely in the healthy puppies, to feed this poor prospect. Not only that, but due to physical mutation reflecting mental mutation (pleiotropy between physical and mental effects) the survival of this puppy could be negative for the survival of the pack. It could, in other words, induce negative social epistasis. Certainly, Chipper was atypical looking for a Springer Spaniel. He was extremely stocky and muscular, in a way that normal Springer Spaniels are not. But the mother was effectively engaging in an act of eugenics, and this is also not uncommon among humans, even in some highly developed civilizations. In fact, it can even be perceived, in some ways, as a relatively slow Life History Strategy action, in which "quantity" is traded for "quality": resources are not wasted on a moribund puppy but are directed towards the healthy puppies so that they can thrive, in a competitive and predictable ecology, to an even greater extent. In fact, there is evidence that poorer families, with poverty indicating that survival is not a certainty, will neglect disabled children relative to their healthy children.[2] In other words, to a certain extent, they reject the "runt of the litter."

Hunter-Gatherers, Pastoralists and Primitive Farmers

Infanticide is known among hunter-gatherers. Among the San tribes of southern Africa, a distinct race also known as the "Bushmen,"[3] infanticide is certainly practiced. According to Canadian anthropologist Nancy Howell, who conducted fieldwork with the San, "infanticide is part of the mother's prerogatives and responsibilities, culturally prescribed for birth defects and

Beach Gardens, FL: Edra Publishing, 2022).

2 C. R. Badcock, *Evolutionary Psychology: A Critical Introduction* (London. Macmillan Higher Education, 2000), 163.

3 See, E. Dutton, *Making Sense of Race* (Whitefish, MT: Washington Summit, 2020).

for one of each pair of twins."[4] The latter form of infanticide is to ensure that at least one of the twins, who are both likely to be born underweight, survives through sufficient access to milk and other resources. High rates of infanticide have also been found among other hunter-gatherer tribes and horticulturalists, such as those in New Guinea and in the Amazon.[5] American anthropologist Napoleon Chagnon (1938–2019), who conducted fieldwork with the Yanomamö of Venezuela, found that defective infants or those whom the mother believed would not thrive, were either smashed against a tree to kill them or they were simply left to slowly die.[6] Among a hunter-gatherer tribe in New Guinea, babies that are clearly disabled are immediately buried alive. These practices also extend to many pastoralist societies. It has been documented that many Native American tribes simply abandoned babies that were born with disabilities.[7] The infanticide specifically of females has been reported in ethnographies on tribal groups in India, with Indian tribal groups tending to be very poor. However, this is not a matter of direct "eugenics" so much as of desiring sons such that they can work on the farm, with daughters being expensive and requiring a dowry in order to marry.[8] In this sense, however, it could be argued that for poorer families, daughters are a drain on their meagre resources. In Uganda there is an ongoing practice whereby disabled infants

4 Quoted in, S. Hrdy, "Fitness Trade-offs in the History and Evolution of Delegated Mothering with Special Reference to Wet-Nursing, Abandonment and Infanticide" in S. Parmiagani and F. vom Saal, (Eds.), *Infanticide and Parental Care* (London: Routledge, 2016).

5 Hrdy, "Fitness Trade-offs in the History and Evolution of Delegated Mothering with Special Reference to Wet-Nursing, Abandonment and Infanticide," *op. cit.*

6 N. Chagnon, *Yanomamö: The Fierce People* (New York: Holt, Rinehart & Winston, 1968).

7 W. Bryan, *Sociopolitical Aspects of Disabilities: The Social Perspectives and Political History of Disabilities and Rehabilitation in the United States* (Springfield, IL: Charles C. Thomas Publisher, 2010), 45.

8 N. P. Das and U. Shah, "Contraceptive Prevalence and Maternal Child Care Status Among Tribals" in R. Nath and P. Lalitendu Jagatdeb, (Eds.), *Tribal Demography in India* (New Delhi: Ashish Publishing House, 1991), 192.

are taken to a witch doctor to be sacrificed.[9]

Where infanticide is not practiced, there is a strong tendency for those who are regarded as physically defective to be severely mistreated, shunned or even accused of witchcraft and executed as a consequence, thus preventing them from passing on their genes. In a modern continuation of what are likely to be ancient practices, disabled children in many Sub-Saharan African countries are at seriously elevated risk of being accused of witchcraft. In Sierra Leone, disabled children accused of witchcraft are forced to drink kerosene. In parts of Ghana, they are chained to trees, denied food, and exposed to the sun in order to exorcise them. Traditional communities in Ghana regard mentally retarded children as "spirit children" who are "not meant for this world" and who have been "sent by evil spirits to bring harm to a family." In the Central African Republic, mentally or physically disabled boys tend to be accused of witchcraft and driven out of their communities, where they are likely to die without assistance. Sometimes the community simply executes them for their "crime." In Togo, children with cerebral palsy are regarded as being possessed by demons and are referred to as "snakes," due to their inability to stand. If found to be "witches," they are either ritualistically drowned or they are taken into the jungle and left there to die. Children with albinism are regarded as sub-human in many rural Sub-Saharan African communities and are mistreated and murdered accordingly.[10] In others, they are perceived as having a special spiritual essence, meaning it is extremely propitious to sacrifice them. Albino children are far more likely to be sacrificed than are Albino adults, because Albino children are easier to kidnap.[11] Thus, these children are prevented from procreating, from being a burden to the community or both, and

9 A. McCrary, "Tracing Tradition: A Case Study of Child Sacrifice in Uganda," *International Letters of Social and Humanistic Sciences*, 41 (2014): 25–38.

10 S. Tsegaye Tesemma and S. A. Coetzee, "Conflicting Discourses on Conceptualizing Children with Disabilities in Africa" *African Disability Rights Yearbook* (Pretoria: Pretoria University Law Press, 2019), 65–68.

11 P. Peek, *Twins in African and Diaspora Cultures: Double Trouble, Twice Blessed* (Indianapolis: Indiana University Press, 2011), 54.

this is, in effect, religiously sanctified. Albinism is, of course, the expression of a mutation, one that is particularly disadvantageous in Africa due to the ease with which Albino skin burns.[12]

Eugenics and the City

Eugenic practices continued with the development of more complex civilizations and of cities. Most obviously, abortion was perfectly acceptable in many pre-Christian populations and, considering its unpleasant nature, was more likely to be practiced out of desperation by poor women who would be likely to have relatively low intelligence and relatively poor moral character.[13] Exposure of unwanted infants was also commonplace in the Ancient World and—unlike in the Christian World—it was perfectly legal. It was particularly common in Athens, for example, after about the fourth century BC, for disabled infants to be left to die. Indeed, a shortage of soldiers in the second century BC in Athens was put down to families preferring to have only one child and duly exposing any others. Poor Athenian families would also sell their children into slavery, which would decrease the likelihood of these children themselves having children.[14]

The most well-known system of eugenics was, of course, that practiced by the city state of Sparta, which was independent from roughly 1000 BC to 200 BC. Its regime of eugenics was established around 620 BC, in the wake of a long and serious guerrilla war campaign against them by the *Helots*, a group whom the Spartans oppressed.[15] All new-born Spartan infants were subject to examination by a special council. If they were judged to be

12 Q. Yang, S. Yi, M. Li et al., "Genetic analyses of oculocutaneous albinism types 1 and 2 with four novel mutations," *BMC Medical Genetics*, 20 (2019): 106.

13 See, K. Kapparis, *Abortion in the Ancient World* (London: Bloomsbury Academic, 2002).

14 Hrdy, "Fitness Trade-offs in the History and Evolution of Delegated Mothering with Special Reference to Wet-Nursing, Abandonment and Infanticide," *op. cit.*

15 M. Burger, *The Shaping of Western Civilization: Volume I From Antiquity to the Mid-Eighteenth Century* (Toronto: University of Toronto Press, 2013), 62.

in any way defective, meaning unfit to be future soldiers, then the parents had to leave them to die via exposure. According to some sources, such infants were thrown into a chasm at the foot of Mount Taygetus. But, exposed, they would either die or, occasionally, be adopted by kindly strangers.[16] In Greek myth, Oedipus was exposed and rescued.[17] At the age of seven, all Spartan boys were placed on an austere military program called the *agoge* in which they were taught the art of war and survival skills. At the age of 12, they were deprived of all clothes, apart from a red cloak, and forced to sleep outside. They were also encouraged to scavenge for and steal food, though they were flogged if they were caught doing the latter.[18] During their time at the *agoge,* these youths were ritualistically beaten, sometimes to death, in front of the altar of Artemis Orthia as part of a ritual of pain-endurance known as *diamastigosis.*[19] In addition, as part of their extended "coming of age" ritual, Spartan youths were sent out to attack the *Helots,* the oppressed slave-class who engaged in much of the physical labour that sustained Spartan society. Roughly 8,000 Spartans ruled over 250,000 Helots, so Spartans were naturally fearful of uprisings,[20] especially considering the uprising of 650–620 BC. The Helots would have defended themselves, meaning that this ritual would have selected for strength, physical prowess and, indeed, cunning and planning and, thus, for intelligence.

Clearly, these kinds of practices would have effectively selected out those with congenitally weak constitutions—and, to some extent, those with low intelligence and low impulse control—and thus those with higher mutational load. Spartans graduated from the *agoge* at 20, after which they could marry. However, they were not allowed to live with their spouse—that is, outside the

16 See N. Kennell, *Spartans: A New History* (Hoboken, NJ: John Wiley & Sons, 2011).

17 S. Basset, "The Exposure of Oedipus," *Classical Review,* 26 (1912): 217.

18 See N. Kennell, *Spartans, op cit.*

19 W. Burkert, *Savage Energies: Lessons of Myth and Ritual in Ancient Greece* (Chicago: University of Chicago Press, 2001), 101.

20 G. Lundskow, *The Sociology of Religion: A Substantive and Transdisciplinary Approach* (Los Angeles: Pine Forge Press, 2008), 152.

barracks—until they were 30, and conjugal visits were strictly regulated. At 30, they became full citizens and could leave the barracks, though they were subject to recall into military service until the age of 60.[21] In this sense, there was a further level of selection: only the most successful soldiers, in a context of hand-to-hand combat, could realistically pass on their genes to a great extent, and only these would survive long enough to live with their wives.

Plato and Eugenics

The Spartans, being highly militaristic, did not focus a great deal on intellectual pursuits and so there are few if any surviving sources penned by Spartan thinkers. The Athenian philosopher Plato (c.428–348 BC), however, was a great admirer of the Spartans, even though Sparta and Athens were sometimes at war. The Ancient Greeks understood that plants and livestock could be bred in order to produce superior strains and in *The Republic* Plato argued that these same processes should be applied to humans. A functioning society needed three key classes—the Guardians, the Soldiers and the Workers or those that, intellectually, were "Gold," "Silver" or "Bronze"—and, in each generation, only the best of these should be permitted to breed in the Gold and Silver classes.

However, Plato realised that this would not be a popular system, so it would have to be hidden from the general public. It would be concealed via a (fixed) lottery to have children where, in reality, those with high lottery numbers would be paired, through the arrangement of meetings, with others who had high lottery numbers and only these "lottery winners" would be permitted to procreate. These winners would be paired-up at special marriage festivals.

> "Had we not better appoint certain festivals at which we will bring together the brides and bridegrooms, and sacrifices will be offered and suitable hymeneal songs composed by our poets: the number of wed-

21 S. Chissanthos, *Warfare in the Ancient World: From the Bronze Age to the Fall of Rome* (Westport, CT: Praeger, 2008), 41.

> dings is a matter which must be left to the discretion of the rulers, whose aim will be to preserve the average of population? There are many other things which they will have to consider, such as the effects of wars and diseases and any similar agencies, in order as far as this is possible to prevent the State from becoming either too large or too small."
>
> "Certainly," he replied. "We shall have to invent some ingenious kind of lots which the less worthy may draw on each occasion of our bringing them together, and then they will accuse their own ill-luck and not the rulers."[22]

Moreover, only the superior "Guardian" children would be allowed to remain in the "Guardian Class," with the inferior ones being relegated to the next class down. Any superior male up to the age of 35 who refused to breed, Plato argued in *The Laws,* should be heavily fined. Aristotle (384–322 BC) criticised and developed Plato's system, arguing that birth-rates should be heavily restricted among the Workers, because otherwise society would dissolve into the violence and disorder to which he seemed to believe that such people were congenitally prone.[23] So, to be clear, academic philosophy was effectively founded by convinced "eugenicists."

Slavery in the United States

After this, we see the collapse of Classical Civilization, and there is little or no overt discussion of eugenics in the West for well over a millennium. However, as we have seen, Western societies—under harsh Darwinian pressures—followed assorted eugenic practices. Only with the Industrial Revolution did thinkers start to notice that mortality was being heavily reduced among the lower classes, people whom they correctly understood to have, on average, lower intelligence, and worse moral character,

22 Plato, *The Republic* (Oxford: Clarendon Press, 1888, Trans. B. Jowett), Book V.

23 D. Galton, "Greek Theories on Eugenics," *Journal of Medical Ethics,* 24 (1998): 263–267.

with GFP being correlated with socioeconomic status at about 0.4.[24]

The only exception is slavery in the southern states of the US, where there were cases of deliberate eugenic breeding.[25] Farmers knew how to breed animals for certain qualities by inter-breeding livestock with these qualities. In at least some cases this technique was consciously applied to black slaves. Virginia was known as the "Negro Raising State" due to the various crude attempts to breed hardier slaves.[26] In his 1845 autobiography of slave life, *Narrative of the Life of Frederick Douglass, An American Slave*, Frederick Douglass (c.1818–1895) recorded his shock that certain slaves were sold to be "breeders." He recalls a young landowner called Covey:

> Mr Covey was a poor man, he was just commencing in life, he was only able to buy one slave, and shocking is the fact, he bought her, as he said, for a breeder. This woman was named Caroline. [..] She was a large, able-bodied woman, about twenty years old [..] After buying her, he hired a married man of Mr Samuel Harrison, to live with him one year; and him he used to fasten up with her every night! The result was, that, at the end of the year, the miserable woman gave birth to twins. At this result Mr Covey seemed to be highly pleased, both with the man and the wretched woman. Such was his joy, and that of his wife, that nothing they could do for Caroline during her confinement was too good, or too hard, to be done. The children were regarded as being quite an addition to his wealth.[27]

24 D. van der Linden, C. Dunkel, and K. V. Petrides, "The General Factor of Personality (GFP) as social effectiveness: Review of the literature," *Personality and Individual Differences,* 101 (2016): 98–105.

25 This section on slavery was originally published in E. Dutton and R. Lynn, *Race and Sport: Evolution and Racial Differences in Sporting Ability* (London: Ulster Institute for Social Research, 2015), 118–119.

26 J. Entine, *Taboo: Why Black Athletes Dominate Sports and Why We're Afraid to Talk About It* (New York: Bantam Books, 2000), 263.

27 F. Douglass, *Narrative of the Life of Frederick Douglass, An American Slave* (Boston, MA: Anti- Slavery Office, 1845), 54–55.

The fact that Mr Covey has a term for his practice would seem to imply that this was not an isolated example. Many accounts by former slaves exist demonstrating that slave masters interfered with the breeding patterns of their slaves, meaning that this kind of eugenics was relatively widely practiced.[28]

In 1988, American sports commentator Jimmy Snyder (1918–1996)—known as "Jimmy the Greek"—was actually fired by CBS Sports for saying that blacks were superior athletes partly because of breeding from the time of slavery: "the black is a better athlete to begin with, because he's been bred that way. Because of his high thighs that go up into his back. And they can jump higher and run faster because of their bigger thighs. This all goes back to the Civil War, when, during the slave trading, the slave owner would breed his big black to his big woman so that he would have a big black kid. That's where it all started."[29]

Modern Advocacy for Eugenics: Benedict Morel

The first person to become concerned about dysgenics—leading, indirectly, to calls for eugenic practices among Europeans—was, seemingly, a French doctor called Benedict Morel (1809–1863). In 1857, Morel published a book in which he pointed out that the infant and child mortality rates in France were clearly in decline.[30] They were in decline, he argued, because various improvements in public health in the preceding 50 years meant that many infants who would have died were now surviving into adulthood. It may sound callous, but there was a clear logic to the argument which followed. Morel regarded the most important qualities as "health," "moral character" (by which he meant, essentially, high Conscientiousness and high Agreeableness), and "intelligence."

Morel was of the view that these qualities were transmitted within families, from parents to children, through a combination

28 See, G. Smithers, *Science, Sexuality and Race in the United States and Australia, 1780s-1890s* (London: Routledge, 2008).

29 G. Solomon, "'Jimmy the Greek' Fired By CBS For His Remarks," *Washington Post* (17th January 1988).

30 B. A. Morel, *Traité des dégenérescenses physiques, intellectuelles et morales de l'espèce humaine* (Paris: Larousse, 1857).

of (what would later be called) genetic and environmental processes. He argued that there was a "degenerate class," an "underclass" of prostitutes, criminals and the desperately poor and that these people—albeit in his subjective experience as a physician and psychiatrist—seemed to have particularly high fertility. So, Morel predicted that these two processes—the reduction in child mortality as a check on the fertility of the "underclass" and the apparent, greater fertility of the underclass—would necessarily lead to the population of France gradually becoming less intelligent. In addition, they would also become—he predicted—more genetically sick.

Francis Galton

The most prominent figure to become concerned about this, and the person who coined the word "eugenics," was English scientist Sir Francis Galton. In his 1869 book *Hereditary Genius,*[31] and in his 1883 book *Inquiries into Human Faculty,*[32] Galton, who was Charles Darwin's cousin, argued that there were three fundamental traits for which humans had long been under selection: health, intelligence, and "moral character," by which he meant Agreeableness and Conscientiousness. Galton averred that high intelligence and good moral character were manifestly central to the maintenance of civilization and, moreover, any civilization was going to be seriously compromised if it was full of people who had chronic poor health.

In *Hereditary Genius,* Galton provided strong evidence that high ability was strongly heritable, as demonstrated by the fact that very high achievers generally seemed to be related to other very high achievers and also as evidenced in people of brilliance who had overcome terrible environments and shone nonetheless. Galton averred that the innovations of high civilization all act to weaken the intensity of natural selection and thus to artificially preserve those of poor health, poor moral character, and low

31 F. Galton, *Hereditary Genius: An Inquiry into Its Laws and Consequences* (London: Macmillan, 1869).

32 F. Galton, *Inquiries into Human Faculty and Its Development* (London: Macmillan, 1883).

intelligence. Not only that but they seem to cause those of the highest social classes—with the highest intelligence and moral character—to eschew having children at all. The result is a kind of "Natural Selection" in reverse, in which society will inevitably degenerate. In Galton's words:

> One of the effects of civilization is to diminish the rigour of the application of the law of Natural Selection. It preserves weakly lives that would have perished in barbarous lands. The sickly children of a wealthy family have a better chance of living and rearing offspring than the stalwart children of a poor one.[33]

This happens, supposed Galton, because large families generally happen by accident with most children of the poor failing to survive. Once all the children of these impulsive poorer people do survive, they will inevitably outbreed the higher classes. The higher classes will want fewer children, anyway, because they will want to dedicate their time and resources to their careers and their interests, meaning they will delay and limit their fertility. The less intelligent and those of worse moral character will be less able to do this. Thus:

> [...] there is a steady check in an old civilisation upon the fertility of the abler classes: the improvident and unambitious are those who chiefly keep up the breed. So the race gradually deteriorates, becoming in each successive generation less fit for a high civilisation.[34]

Galton averred that this was a very serious problem which needed to be solved. Galton even conceived of a utopia called "Kantsaywhere," which would be run by a "Eugenic College." It would preside over a system whereby all citizens would take examinations of strength, aesthetics, health, and would be tested for a healthy and intelligent genealogy. Only those who passed the exam would be allowed to have children and the number they can have would vary according to their mark, from 1 (as many

33 F. Galton, "Hereditary talent and character," *MacMillan's Magazine*, 12 (June-August 1865): 314.

34 F. Galton, *Hereditary Genius, op. cit.*, 414.

children as they wish) to 4 (one child). Those given a "5" would not be given a licence to have children. Galton noted that this would have to be carried out in an authoritarian state and that a further problem was that there was no guarantee that it would be the best of these selected breeders who had the most children.[35] So, how could these problems be solved?

Eugenics as an Academic Discipline

Firstly, argued Galton, an academic discipline of "Eugenics" needed to be developed so that the best minds could be set to solve the problem. This is precisely what happened. In 1904, Galton set up the "Eugenics Record Office" in London. In 1907, this was renamed the "Galton Eugenics Laboratory" and it soon became part of University College London (UCL). The laboratory was directed by Karl Pearson (1857–1936), Galton's protégé and biographer, who was Professor of Applied Mathematics at UCL. Galton, by the way, invented correlational analysis, which was further developed by Pearson, a hugely important contribution to statistics.[36]

When the unmarried Galton died in 1911, he left sufficient funds to establish a chair in "National Eugenics," with Pearson, therefore, becoming "Galton Professor of Eugenics," at University College London. In 1945, this post was renamed "Galton Professor of Human Genetics." In 1925, Pearson set up, and became the first editor of, the academic journal *Annals of Eugenics,* which was later renamed *Annals of Human Genetics.* In 1933, Pearson was succeeded as Galton Professor of Eugenics by Sir Ronald Fisher (1890–1962).[37] We will explore the development of "eugenics" within academia in more detail below.

35 R. Lynn, *Eugenics: A Reassessment* (Westport, CT: Praeger, 2001), 15.

36 J. Stanton, "Galton, Pearson, and the Peas: A Brief History of Linear Regression for Statistics Instructors," *Journal of Statistics Education,* 9 (2017): 3.

37 Lynn, *Eugenics, op. cit.*

Eugenics Promoted to the Public

Secondly, argued Galton, the public had to be persuaded of the importance of eugenics. Towards the end of his life, he suggested that this should be via the establishment of a kind of ersatz religion:

> It must be introduced into the national conscience, like a new religion. It has, indeed, strong claims to become an orthodox religious tenet of the future, for eugenics co-operate with the workings of nature by securing that humanity shall be represented by the fittest races. What nature does blindly, slowly, and ruthlessly, man may do providently, quickly, and kindly. As it lies within his power, so it becomes his duty to work in that direction.[38]

Eugenics did, indeed, become popular among the educated classes up until World War II. For Galton, the public needed to be persuaded that, in eugenic terms, there were definite "desirables," "passables", and "undesirables." The "desirables"—which included not merely the higher classes and the highly intelligent but also groups such as "contented labourers"—needed to be induced, such as via financial incentives, to have large numbers of children. These inducements would not be offered to those who were regarded merely as "passables." This was "positive eugenics." As for the congenital criminals and psychopaths who made up the "undesirables"—and, in Galton's observation, seemed to have high fertility—then "negative eugenics" was required. The "underclass" needed to be forcibly sterilized.[39] Galton was also in favour of modest levels of highly intelligent immigration, of "desirable refugees," and he highlighted the French Huguenots as having been examples of these.[40]

In terms of persuading the public of the importance of eu-

38 F. Galton, "Eugenics: Its Definition, Scope and Aims," *American Journal of Sociology,* 10 (1904): 1.

39 F. Galton, *Memories of My Life* (London: Methuen and Co., 1909), 322.

40 F. Galton, *Hereditary Genius, op cit.*, 360

genics, the "Eugenics Education Society" was set up in 1907, at around the time that people were starting to become concerned about the birth-rate, which had fallen by 37 per cent between 1870 and 1910.[41] Fifty per cent of the society's early members were women, especially teachers and social workers.[42] The Eugenics Education Society was renamed the "Eugenics Society" in 1924 and by the 1930s over 40 per cent of its members were female.[43] Until 1968, the society published a journal called the *Eugenics Review.* Some of the most eminent scientists of the first half of the twentieth century were active in the Eugenics Society including Sir Ronald Fisher, the economist John Maynard Keynes (1883–1946), the biologist Sir Julian Huxley (1887–1975) and the anthropologist George Pitt-Rivers, whom we discussed earlier. As one historian has put it, the membership of the Eugenics Society is "a list of brilliant names; one can be in no doubt of the status of the Eugenics Society: it could count among its own members, or co-opt when necessary, some of the most distinguished men and women of its age."[44]

Arthur Balfour (1848–1930), who was Conservative Prime Minister between 1902 and 1905, was an enthusiastic member of the Eugenics Society and persuaded a donor to help endow a chair in genetics at Cambridge University in order to further the study of eugenics.[45] By 1911, it had become clear from census returns and other reports that, in essence, those of low socioeconomic status were heavily outbreeding those of high socioeconomic status, with it being correctly assumed that the related

41 A. T. Allen, "Feminism and Eugenics in Germany and Britain, 1900–1940: A Comparative Perspective," *German Studies Review,* 23 (2000): 477–505.

42 L. Hall, Sex, *Gender and Social Change in Britain Since 1880* (London: Bloomsbury, 2017), 59.

43 D. Childs, *Modernism and Eugenics: Woolf, Eliot, Yeats, and the Culture of Degeneration* (Cambridge: Cambridge University Press, 2001), 8.

44 P. Mazumdar, *Eugenics, Human Genetics and Human Failings: The Eugenics Society, Its Sources and Its Critics in Britain* (London: Routledge, 2005), 36.

45 T. Hamilton, *Arthur Balfour's Ghosts: An Edwardian Elite and the Riddle of the Cross-Correspondence Automatic Writings* (Luton: Andrews UK Ltd, 2018).

qualities, such as intelligence, were largely genetic.[46] Balfour referred to eugenics as "the greatest and most pressing necessity of the age" at the Eugenics Congress in London in summer 1912.[47] Neville Chamberlain (1869–1940), who was Conservative Prime Minister from 1937 to 1940, was an active member of the Eugenics Society branch in Birmingham.[48]

The influence of the Eugenics Society is evident in the fact that it had branches in most major towns, organised summer schools at provincial colleges,[49] gave talks to civic groups such as the Rotarians[50] and the Women's Institute,[51] presented before crowded public meetings in such towns,[52] and was supported by the president of the British Medical Association.[53] The Eugenics Society cooperated with charities for the disabled, such as to pressure the government for the compulsory sexual segregation of "mental defectives."[54] In many respects, eugenics was very much the "current thing" of its time.[55] In the 1930s, William Beveridge (1879–1963), the economist whose later report helped to found Britain's "Welfare State" and who is regard as the architect of this and of the National Health Service in particular,[56] stressed that it was important to encourage the fertility of "good" rather than

46 R. L. Soloway, *Demography and Degeneration: Eugenics and the Declining Birthrate in Twentieth-Century Britain* (Chapel Hill, NC: University of North Carolina Press, 1995), 16.

47 *Dundee Courier*, "Balfour and Eugenics" (25th July 1912).

48 G. R. Searle, *Eugenics and Politics in Britain, 1900–1914* (Leyden: Nordhoof International Publishing, 1976), 13.

49 *Westminster Gazette*, "Summer School of Eugenics" (1st July 1920).

50 *Beds and Herts Pictorial*, "Principles of Eugenics" (17th October 1933).

51 R. Youngson and I. Schott, *Medical Blunders: Amazing True Stories of Mad, Bad, and Dangerous Doctors* (New York: New York University Press, 1998), 362.

52 *Reading Observer*, "Dean Inge On Eugenics" (29th January 1921).

53 *The Cornish Telegraph*, "Sir J. Barr on Eugenics" (2nd January 1913).

54 *Hartlepool Northern Daily Mail*, "Eugenics Society" (14th June 1924).

55 M. Read, "What is 'The Current Thing'?" *The Slate* (9th May 2022), https://slate.com/technology/2022/05/current-thing-meme-twitter-elon-musk-marc-andreessen.html.

56 D. Pilgrim, *Key Concepts in Mental Health* (London: Sage, 2017).

merely "ordinary" people.[57] Beveridge was not a member of the Eugenics Society, but he was evidently a sympathiser,[58] giving many presentations to them.[59] Britain's "Welfare State," then, as well as "Our NHS," was designed by a "eugenicist."

The influence of eugenicists, and the widespread acceptance of their ideas, is also evident in just how shameful it was to have a relative who had any kind of genetic disorder or even a disorder that might be genetic. Such a family was perceived to be "tainted" with "bad blood."[60] In the British royal family, the mildly mentally retarded and epileptic Prince Johnny (1905–1919)—the youngest of the children of King George V (1865–1936)—was secluded from the public; his problems not reported. They only came to light after his death.[61] It was revealed in 1987 that sisters Nerissa (1919–1986) and Katherine Bowes-Lyon (1926–2014), maternal cousins of Queen Elizabeth II (1926–2022), had been placed in Earlswood, a psychiatric hospital in Surrey, in 1941. Nerissa spent the rest of her life there. Both sisters were severely mentally retarded. *Burke's Peerage* had, as of its 1963 edition, listed Nerissa as dead since 1940 and Katherine as dead since 1961, with specific dates of death provided by their mother.[62]

Eugenics and the Left

Many people on the left of British politics, however, were open eugenicists, with most of the criticism coming from conserva-

57 Mazumdar, *Eugenics, Human Genetics and Human Failings, op. cit.,* 34.

58 Mazumdar, *Eugenics, Human Genetics and Human Failings, op. cit.,* 36.

59 *Aberdeen Evening Express,* "Urgent Problem of the Birthrate" (16th February 1943).

60 M. Pernick, *The Black Stork: Eugenics and the Death of "Defective" Babies in American Medicine and Motion Pictures since 1915* (Oxford: Oxford University Press, 1996), 52.

61 P. Levine, *Eugenics: A Very Short Introduction* (Oxford: Oxford University Press, 2017), 81–82. *Aberdeen Press and Journal,* "Sudden Death of Prince John" (20th January 1919).

62 H. Coke, "Behind The Crown: The true story of the Queen's cousins, Nerissa and Katherine Bowes-Lyon," *Tatler* (19th November 2020).

tives, as we will see below. These leftists tended to claim to be motivated by a desire to reduce future suffering, both of sick individuals and in society as a whole. The runaway individualism which has led to arguments such as "all humans are of equal value, including the genetically seriously unfit, no matter the cost they impose on society" had not yet been reached in this Britain. This was also a Britain in which poverty and mortality salience were much more widespread and in which relatively little could be done to alleviate the suffering caused by serious illness. In addition, the relationship between leftism and Neuroticism, discussed above, could mean that leftists were actually more fearful of the dangers of the underclass, in terms of violence and disease, than were conservatives. Irish playwright George Bernard Shaw (1856–1950) wrote: "The only fundamental and possible socialism is the socialisation of the selective breeding of man."[63]

Family Planning

Family-planning pioneer Marie Stopes (1880–1958) was motivated by a desire to limit the fertility of the "undesirables." Stopes herself proclaimed in 1919, "more children from the fit, less from the unfit, that is the chief issue of birth control."[64] Her American counterpart, Margaret Sanger (1879–1966), was also supportive of eugenics. It directly informed her need to promote "planned parenthood."

> Eugenics seems to me to be valuable in its critical and diagnostic aspects, in emphasizing the danger of irresponsible and uncontrolled fertility of the "unfit" and the feeble-minded establishing a progressive unbalance in human society and lowering the birth-rate among the "fit."[65]

"Planned Parenthood," then, and attempts to provide contracep-

63 J. Freedland, "Eugenics and the master race of the left," *The Guardian* (30th August 1997).

64 R. Weikart, *From Darwin to Hitler: Evolutionary Ethics, Eugenics and Racism in Germany* (New York: Springer, 2016), 135.

65 M. Sanger, *The Pivot of Civilization* (New York: Bentano's, 1922), 130.

tion for women, were motivated by eugenics.

Social Work

The profession of social worker had developed by the early twentieth century and leading social workers, including pioneers of the profession such as Americans Mary Richmond (1866–1928)[66] and Jane Addams (1860–1935),[67] as well as ordinary practitioners, were strong supporters of eugenics, seeing social work as a means of segregating the fit—who could benefit from the advice and assistance of social workers—from the unfit, and so improve society. They especially focused on young, non-WASP women: "Social workers scientifically assessed adolescent women via mental testing for feeblemindedness, arranged physical examinations to ascertain evidence of immorality, and then recommended segregation and institutionalization on the basis of their expert diagnoses."[68]

The Literary World

The literary world was also replete with eugenicists. English novelist H. G. Wells (1866–1946) enthused that eugenics allowed the removal "of detrimental types and characteristics" and the "fostering of desirable types" in their place.[69] In his 1905 novel *A Modern Utopia,* Wells conceived of a world where you could not have children until you'd worked a number of years and you had no debts. Moreover:

66 K. Wright, K. Carr, and B. Akin, "The Whitewashing of Social Work History: How Dismantling Racism in Social Work Education Begins With an Equitable History of the Profession," *Advances in Social Work,* 21 (2021): 274–297.

67 G. O'Brien, *Eugenics, Genetics, and Disability in Historical and Contemporary Perspective: Implications for the Social Work Profession* (Oxford: Oxford University Press, 2023), 57.

68 A. C. Kennedy, "Eugenics, "Degenerate Girls," and Social Workers During the Progressive Era," *Affilia,* 23 (2008): 22–37.

69 J. Freedland, "Eugenics and the master race of the left," *The Guardian* (30th August 1997).

> There remain idiots and lunatics,[70] perverse and incompetent persons, people of weak character who become drunkards, drug takers, and the like, and persons tainted with certain foul and transmissible diseases. All of these people spoil the world for others. They may become parents, and with most of them there is nothing to be done, but to seclude them from the great body of the population. [71]

This would be accomplished, in Wells' utopia, by deporting them to a heavily guarded island. Other literary figures concurred. Novelist Aldous Huxley (1894–1963), who penned the novel *A Brave New World* on the subject of a eugenics-based dystopia,[72] wrote in 1927 that, "in the future we envisage that eugenics will be practiced in order to improve the human breed."[73] Essayist and novelist D. H. Lawrence (1885–1930) strongly advocated "negative eugenics," removing the unfit rather than simply encouraging the fit to breed:

> If I had my way, I would build a lethal chamber as big as Crystal Palace, with a military band playing softly and a Cinematograph working brightly; then I'd go out in the back streets and main streets and bring them all in, all the sick, the halt and the maimed; I would lead them gently and they would smile me a weary thanks, and the band would softly bubble out the "Hallelujah Chorus."[74]

70 In Victorian parlance, an "Idiot" was someone who was mentally retarded whereas a "Lunatic" was a person who had once been normal but who had "lost his mind." D. Wright, *Mental Disability in Victorian England: The Earlswood Asylum 1847–1901* (Oxford: Clarendon Press, 2001), 10.

71 H. G. Wells, *A Modern Utopia* (London: Chapman and Hall, 1905), 141–142.

72 A. Huxley, *A Brave New World* (London: Chatto and Windus, 1932).

73 A. Huxley, "The Future of the Past: How Humans Will Eventually Be Classified By Ability Rather Than Prestige," *Vanity Fair* (September 1927).

74 Quoted in: K. Malik, *The Meaning of Race: Race, History and Culture in Western Society* (New York: New York University Press, 1996), 114.

Leftist Politicians

The left-wing political scene also included many eugenicists. By the 1930s, the Eugenics Society rivalled the Fabian Society as the fashionable hang-out for London leftists. Labour Member of Parliament Ellen Wilkinson (1891–1947) proposed that the society form a "committee of Labour sympathisers." Leftist organs such as *The New Statesman* and *The Manchester Guardian* ran many pro-eugenics pieces.[75]

Labour Member of Parliament Major Archibald Church (1886–1954), who was General Secretary of the National Union of Scientific Workers, proposed a bill in 1931 for the voluntary sterilisation of mental "defectives." Continental Social Democrats enthusiastically espoused, and unlike in Britain *passed*, sterilisation laws. Swedish Social Democrat Member of Parliament Karl Johan Olsson (1893–1980) proclaimed that, "I think it is better to go a little too far than to risk bringing unfit and inferior offspring into the world."[76] Socialist philosopher and sometime Labour Party parliamentary candidate[77] Bertrand Russell (1872–1970) proclaimed:

> There can be no doubt that the civilization produced by the white races has this singular characteristic, that in proportion as men and women absorb it, they become sterile. The most civilized are the most sterile, the least civilized are the most fertile, and between the two there is a continuous gradation. At present the most intelligent sections of the Western nations are dying out.[78]

Karl Pearson, whom we met earlier, was also a convinced social-

75 Freedland, "Eugenics and the master race of the left," *op. cit.*

76 M. Conroy, *Nazi Eugenics: Precursors, Policy, Aftermath* (New York: Columbia University Press, 2017).

77 A. C. Grayling, *Russell: A Very Short Introduction* (Oxford: Oxford University Press, 2002), 20.

78 B. Russell, *The Conquest of Happiness* (London: W. W. Norton, 1930), 234.

ist.[79] Indeed, recent research from the US has found that liberals actually tend to be slightly more in favour of eugenic policies—such as gene editing for disease and even increasing intelligence through this process—than conservatives.[80]

Liberals and Eugenics

Not long after Galton began advocating for eugenics, he recruited some very prominent supporters. After reading *Hereditary Genius,* Charles Darwin (1809–1882) wrote to Galton: "You have made a convert."[81] In other words, one of the greatest geniuses and most important scientists in world history, whose theory of evolution underpins modern biology, was a "eugenicist." In his 1871 book *The Descent of Man,* Darwin spelt out the problem of dysgenics and the lamentable effects that it would likely have:

> With savages, the weak in body or mind are soon eliminated; and those that survive commonly exhibit a vigorous state of health. We civilised men, on the other hand, do our utmost to check the process of elimination; we build asylums for the imbecile, the maimed, and the sick; we institute poor-laws; and our medical men exert their utmost skill to save the life of every one to the last moment. There is reason to believe that vaccination has preserved thousands, who from a weak constitution would formerly have succumbed to small-pox. Thus the weak members of civilised societies propagate their kind. No one who has attended to the breeding of domestic animals will doubt that this must be highly injurious to the race of man. It is surprising how soon a want of care, or care wrongly directed, leads to the degeneration of a domestic race; but excepting in the case of man himself, hardly any one is so ignorant as to allow his worst

79 B. Semmel, "Karl Pearson: Socialist and Darwinist," *British Journal of Sociology,* 9 (1958): 111–125.

80 E. Kirkegaard, "Who supports biotech eugenics?" *Just Emil Kirkegaard Things* (12th October 2022).

81 Lynn, *Eugenics, op. cit.*

> animals to breed.[82]

Darwin was a "Whig," later known as the Liberal Party,[83] which was the more left-wing party prior to the rise of the Labour Party which displaced the Liberal Party in the 1922 General Election.[84] The Liberal Prime Minister who lost this election, David Lloyd George (1863–1945), expressed concern, in 1918, about running an "A1 empire with a C3" population, arguing that eugenic measures were needed to solve this.[85] A1 and C3 are categories of fitness accorded in army examinations.

In 1873, sociologist and journalist Herbert Spencer (1820–1903), who coined the term "the survival of the fittest,"[86] argued, in the Liberal journal *The Contemporary Review,* that the rise of charities that assisted the genetically unhealthy and the congenitally feeble-minded to survive, was a serious problem, based on Galton's arguments.

> Fostering the good-for-nothing at the expense of the good is an extreme cruelty. It is a deliberate storing up of miseries for future generations. There is no greater curse to posterity than that of bequeathing them an increasing population of imbeciles.[87]

Spencer was an active supporter of the Liberal Party.[88] It was common for eugenics advocates to stress just how damaging to the genetic health of the population assisting the poor was likely to be. In effect, they averred that: "Charity went against nature.

82 C. Darwin, *The Descent of Man* (London: John Murray, 1871), 501.

83 G. West, *Charles Darwin: The Fragmentary Man* (London: Routledge, 1937).

84 R. Douglas, *Liberals: The History of the Liberal and Liberal Democrat Parties* (London A. & C. Black, 2005), 324.

85 N. Shasore, *Designs on Democracy: Architecture and the Public in Interwar London* (Oxford: Oxford University Press, 2022), 324.

86 I. W. Howerth, "Natural Selection and the Survival of the Fittest," *Scientific Monthly,* 5: (1917): 253–257.

87 H. Spencer, "The Study of Sociology," *Contemporary Review,* 22 (1873): 338.

88 M. Francis, *Herbert Spencer and the Invention of Modern Life* (London: Routledge, 2014).

Eugenics, by contrast, might assist nature."[89] For some of these figures, their "eugenics" was motivated by a desire to reduce the suffering of future generations. William Beveridge, the eugenics-supporter whom we met earlier, was briefly a Liberal Member of Parliament before becoming a Liberal Peer.[90]

Eugenics, Traditionalism, and the Radical Right

It was traditionalists—those on the right—who were more likely to oppose eugenics during this period. For example, conservative writer G. K. Chesterton (1874–1936) wrote his book *Eugenics and Other Evils* in 1922, condemning eugenics because it was supposedly non-traditional, involved a supposed lack of compassion, involved making harsh calculations, and did not leave things up to God, meaning it undermined the sacred:

> To be consistent, they ought to put up statues to the men who deserted their loves because of bodily misfortunes; with inscriptions celebrating the good Eugenicist who, on his fiancée falling off a bicycle, nobly refused to marry her [...] Briefly, we may say that while many moral systems have set restraints on sex almost as severe as any Eugenist [*sic.*] could set, they have almost always had the character of securing the fidelity of the two sexes to each other, and leaving the rest to God. To introduce an ethic which makes that fidelity or infidelity vary with some calculation about heredity is that rarest of all things, a revolution that has not happened before.[91]

In other words, they opposed it as a broader means of signalling their traditional religiousness in a broadly conservative society. In the above quote, we can see a straw man argument (eugenicists

89 A. Richardson, *Love and Eugenics in the Late Nineteenth Century: Rational Reproduction and the New Woman* (Oxford: Oxford University Press, 2003), 64.

90 F. Boyd, *British Politics in Transition, 1945–63* (London: Praeger, 1964), 46.

91 G. K. Chesterton, *Eugenics and Other Evils* (London: Cassell and Company, 1922), 7–8.

wanted to prevent congenital disability, not abandon the injured) and simply an appeal to religious tradition. Conservatives such as Chesterton opposed eugenics simply because it was something new, associated with other relative novelties such as Darwinian Theory, whereas progressives would regard it as inherently good precisely because it was something new; with liberalism tending to be strongly associated with Openness.[92] Many other fallacious methods can be found in Chesterton's polemic against eugenics, such as his referring to proponents possessing a "creepy simplicity of mind."[93] Clearly, this is an "appeal to insult."

But, even so, there were some eugenics supporters among those who were highly traditionalist. In 1920, the Very Rev'd William Inge (1860–1954), Dean of St Paul's Cathedral, who notably was opposed to democracy,[94] presented a Christian case for eugenics in his book *Outspoken Essays.* He maintained that Christians are primarily concerned with the intrinsic quality of people, as are eugenicists, and the Sermon on the Mount implies that humans are animals that can be worse or better bred: "A good tree cannot bring forth evil fruit."[95] Inge referred to Chesterton as an "irrationalist,"[96] asserting that: "Eugenicists believe that unless civilisation is guided on scientific principles it must come to ruin [...] we stand by scientific as against emotional or sentimentalist ethics."[97] He averred that "a certificate of sound health should be one of the things insisted on before marriage," before an audience of female students in 1910.[98] This was, of course, in a context in which pre-marital sex and illegitimacy were strongly taboo. In Britain, in the 1930s, children who were known to be illegitimate might find themselves shunned in the

92 J.-E. De Neve, "Personality, Childhood Experience, and Political Ideology," *Political Psychology,* 36 (2015): 55–73.

93 Chesterton, *Eugenics and Other Evils, op. cit.*, 8.

94 W. Inge, *The End of An Age and Other Essays* (London: William Putnam's Sons, 1948).

95 W. Inge, *Outspoken Essays* (London: Longman, Green and Co., 1920), 274.

96 Inge, *Outspoken Essays, op. cit.,* 255.

97 Inge, *Outspoken Essays, op. cit.,* 256.

98 *Dundee Evening Telegraph,* "Eugenics and Love" (10th March 1910).

playground or refused service in shops. Many people saw them as being from "such a morally weak bloodline that they could corrupt others just by being in their presence."[99] In 1911, the chairman of the Belfast Branch of the Eugenics Education Society was the Church of Ireland Bishop of Down and Connor, the Rt. Rev'd Charles D'Arcy (1859–1938).[100] D'Arcy eventually became Archbishop of Armagh and, so, primate of the Church of Ireland.[101] In 1912, the Very Rev'd William Taylor Sumner (1867–1933), Dean of the Episcopal Cathedral in Chicago, announced that he wouldn't marry couples unless they presented him with a medical certificate proving that they were physically and mentally healthy.[102]

Support among mainstream conservatives was substantial, as evidenced in the involvement of three Conservative Prime Ministers already noted: Balfour, Chamberlain, and Churchill. Galton himself favoured the Conservative Party.[103] However, eugenics also had support on the "radical right." Philosopher Anthony Ludovici (1882–1971) went so far as to espouse state-directed infanticide in pursuit of eugenics.[104] Ludovici was a member of the radical right-wing and Pro-Nazi "Right Club" in the 1930s, which was led by Conservative Member of Parliament Captain Archibald Ramsay (1894–1955). George Pitt-Rivers, whom we met earlier, was also a Right Club member,[105] in addition to being

99 S. Danczuk and M. Baker, *Smile for the Camera: The Double Life of Cyril Smith* (London: Biteback Publishing, 2014), 4.

100 *The Irish News*, "Belfast Branch, Eugenics Education Society" (25th November 1911).

101 N. Richardson, *According to Their Lights: Stories of Irishmen in the British Army, Easter 1916* (Dublin: Gill and Macmillan, 2015).

102 P. Lombardo, *Three Generations, No Imbeciles: Eugenics, the Supreme Court, and Buck v. Bell* (Baltimore, MD: Johns Hopkins University Press, 2022).

103 R. S. Cowan, *Sir Francis Galton and the Study of Heredity in the Nineteenth Century* (New York: Garland Publishing, 1985), 56.

104 D. Stone, "Race in British Eugenics," *European History Quarterly*, 3 (2001): 397–425.

105 G. C. Webber, *The Ideology of the British Right, 1918–1939* (London: Routledge, 2015).

a member of the council of the Eugenics Society.[106] In 1940, with Britain at war with Nazi Germany, Ludovici's home was raided by the police and he was interrogated, though he was not interned under wartime regulations as a danger to the state, as Captain Ramsay was.[107] Pitt-Rivers' home was also raided in 1940. He was far more openly pro-Hitler than Ludovici.[108] Pitt-Rivers was arrested and interned between 1940 and 1941.[109]

Academic Eugenics and Eugenic Laws

After Galton's death in 1911, "eugenics research" continued to grow as an academic discipline and also began to heavily influence public policy in many different countries. Such people were convinced that, with the right measures, civilization perhaps need never collapse. As American eugenicist Lothrop Stoddard (1883–1950) spoke for them when he wrote: "Civilizations, unlike living organisms, have no appointed cycle of life and death. Given a high-type stock producing an adequate quota of superior individuals, and a civilization might be immortal."[110]

The United Kingdom

Charles Spearman, who came up with the concept of *g* and was Professor of the Philosophy of Mind at University College London, was a member of the Eugenics Society and published in its journal.[111] Spearman argued in 1912—the year after he was made a professor—that only those who scored reasonably highly on

106 R. Griffiths, *Patriotism Perverted: Captain Ramsay, the Right Club, and British Anti-Semitism, 1939–1940* (London: Faber and Faber, 2016).

107 R. Griffiths, *What Did You Do During the War? The Last Throes of the British Pro-Nazi Right* (London: Routledge, 2017), 327–328.

108 B. W. Hart, *George Pitt-Rivers and the Nazis* (London: Bloomsbury, 2015), 1–2.

109 Griffiths, *What Did You Do During the War? op. cit.*, 263–264.

110 L. Stoddard, *The Revolt Against Civilization: The Menace of the Underman* (New York: Charles Scribner's Sons, 1922), 11.

111 C. Spearman, "The Heredity of Abilities," *Eugenics Review*, 6: (1914–15): 219–237.

g should be allowed to vote or have children.[112] Spearman also invented "factor analysis," a hugely important contribution to statistics.[113] World War I saw particular concern about dysgenics, both in the UK and the US, because it was observed that the officers were more likely to be killed, as they tended to be of higher intelligence, as were brave, self-sacrificing men of all ranks who had pro-social personalities. This problem was set out in a book published during the war by American philosopher and chancellor of Stanford University[114] David Starr Jordan (1851–1931) entitled *War and the Breed.*[115] Concern was raised about this in the UK by chairman of the Eugenics Society Major Leonard Darwin (1850–1943), son of Charles Darwin, who observed in a study published in 1916 that casualties among the officers were double those among the troops,[116] as they were leading groups into battle and, thus, taking greater risks. As one historian has summarised the contemporary view: "The Great War was a eugenics nightmare. It allegedly destroyed the finest physical, mental, and social stock in the country and seriously disrupted family life and selective reproduction."[117] Thus, the popularity of eugenics was rejuvenated after World War I.[118]

Sir Ronald Fisher, Professor of Eugenics at UCL, argued in his book *The Genetical Theory of Natural Selection* that all mature societies start to become dysgenic because they start to protect people against the rigours of Natural Selection and also because, as their needs are increasingly met, they become increasingly

112 Lynn, *Eugenics, op cit.*, 22.

113 D. J. Bartholomew, "Spearman and the origin and development of factor analysis," *British Journal of Mathematical and Statistical Psychology*, 48 (1995): 211–220.

114 *Linlithgowshire Gazette*, "Eugenics of War" (1st August 1913).

115 J. S. Jordan, *War and the Breed: The Relation of War to the Downfall of Nations* (Boston, MA: The Beacon Press, 1915).

116 L. Darwin, "On the Statistical Enquiries Needed After the War in Connection with Eugenics," *Journal of the Royal Statistical Society*, 79 (1916): 155–188.

117 R. L. Soloway, *Demography and Degeneration: Eugenics and the Declining Birthrate in Twentieth-Century Britain* (Chapel Hill, NC: University of North Carolina Press, 1995), 86.

118 Levine, *Eugenics, op. cit.*, 5

concerned about the less well-off (the less intelligent and those of poor genetic health), and they have the resources to assist them. Fisher demonstrated that this could be seen in Greek and Islamic civilization and was the ultimate reason for their collapse, due to declining intelligence. Fisher suggested dealing with this problem by encouraging wealthier people to have children which could be done by making child allowance a proportion of income. This is because Fisher believed that wealthier people eschewed children due to expense. He also argued that moral pressure should be put upon them.[119]

J. B. S. Haldane (1892–1964), who succeeded Fisher as Galton Professor of Eugenics at University College London, argued that a sperm bank should be set up for the genetic elite in order to improve the quality of the national stock.[120] Haldane was also a convinced Marxist who espoused eugenics in the pages of the *Daily Worker* in 1949, writing: "The dogma of human equality is no part of Communism [...] 'from each according to his ability, to each according to his needs' would be nonsense if abilities were equal."[121] In 1959, Sir Peter Medawar (1915–1992), Professor of Zoology at University College London, who was a member of the socialist Labour Party in the 1930s,[122] argued in *The Future of Man,* a BBC Reith Lecture, that British intelligence was clearly declining for genetic reasons and that there needed to be a "humane solution" to the problem.[123] Medawar received the Nobel Prize for Science the following year.[124] Nobel Prize-winning discoverer of the DNA double helix Francis Crick (1916–2004), Fellow of the Salk Institute for Biological Studies in San Diego,

119 R. Fisher, *The Genetical Theory of Natural Selection* (Oxford: Clarendon Press, 1930).

120 Lynn, *Eugenics, op. cit.,* 21.

121 D. Paul, "Eugenics and the Left," *Journal of the History of Ideas,* 45 (1984): 567–590.

122 Anon, "Lady Jean Medawar," *British Medical Journal,* 330 (2005): 1392.

123 P. Medawar, *The Future of Man: The BBC Reith Lectures, 1959* (London: Basic Books, 1959).

124 A. Silverstein, "The curious case of the 1960 Nobel Prize to Burnet and Medawar," *Immunology,* 147 (2016): 269–274.

California, averred, at a conference in 1963, that there should be a tax on children such that the poor—who would tend to be of low intelligence—might be discouraged from having any.[125] Crick was an undergraduate at UCL and was eventually appointed an honorary Fellow there.

The closest Britain came to passing a eugenic law was the 1913 "Mental Deficiency Act." This made it compulsory to hold mentally retarded persons in single-sex institutions in order to avoid unwanted pregnancies. While he was Home Secretary, Winston Churchill (1874–1965) became interested in eugenics. In 1908, a Royal Commission on the feeble-minded, set up by Churchill four years earlier,[126] had recommended the sterilization of Britain's 120,000 "mental deficients" to prevent the spread of "undesirable traits." In 1910, Churchill circulated to his Cabinet colleagues an article by English psychiatrist A. F. Tredgold (1870–1952), one of the advisers to the Royal Commission. It was called "The Feeble-Minded: A Social Danger."[127] Written in 1909, the article gave, in the words of Churchill, "a concise, and, I am afraid not exaggerated statement of the serious problems to be faced." Churchill added: "The Government is pledged to legislation, and a Bill is being drafted to carry out the recommendations of the Royal Commission."[128] Churchill asked his officials to consider a law sterilising the "mentally unfit" and preventing them from marrying, but it came to nothing.[129] In 1912, Churchill, along with Arthur Balfour, was the vice-president of the International Eugenics Congress, which was held in London.[130]

125 Lynn, *Eugenics, op cit.*, 22.

126 See, E. Fry, F. Galton, W. R. Inge, et al., *The Problem of the Feeble-Minded: An Abstract on the Report of the Royal Commission on the Care and Control of the Feeble-Minded* (London: P. King & Son, 1909).

127 A. F. Tredgold, "The Feeble-Minded: A Social Danger," *Eugenics Review*, 1 (1909): 97–104.

128 M. Gilbert, "Churchill and Eugenics," *International Churchill Society* (31st May 2019).

129 R. Langworth, *Winston Churchill, Myth and Reality: What He Actually Did and Said* (Jefferson, NC: McFarland and Company, 2017), 201.

130 D. MacKenzie, "Eugenics in Britain," *Social Studies of Science*, 6 (1976): 499–532.

The United States

In the United States, eugenics quickly attracted many prominent advocates including biologist Charles Davenport (1866–1944), Professor of Zoology at Harvard, who founded the "American Breeders Association" in 1903, to apply agricultural breeding techniques to humans. He established the "Eugenics Record Office" in 1910, and spent his career looking into the heritability of assorted pathologies, such as alcoholism.[131] Alcoholism appears to be 0.5 heritable.[132] Davenport's book *Heredity in Relation to Eugenics* was published in 1911 and was a textbook on many medical courses until the 1940s.[133] Psychologist Lewis Terman (1877–1956), Professor of Educational Psychology at Stanford and well-known for his research into intelligence and personality, espoused a eugenics program,[134] as did English psychologist William MacDougall (1871–1938), who made many discoveries about the nature of instinct and who worked at Duke University in Durham, North Carolina. Harry Laughlin (1880–1943), who had a doctorate in cytology from Princeton, was Superintendent of the Eugenics Record Office, which was supported by, among others, the businessman John Harvey Kellogg (1852–1943) and the Carnegie Institute.[135]

As in Britain, many prominent Americans supported eugenic policies in the first half of the twentieth century including President William Howard Taft (1857–1930),[136] President Theodore

131 B. Kimmelman, "The American Breeders Association: Genetics and Eugenics in an Agricultural Context, 1903–1913," *Social Studies of Science*, 13 (1983): 163–204.

132 B. Verhulst, M. C. Neale, and K. S. Kendler, "The heritability of alcohol use disorders: a meta-analysis of twin and adoption studies," *Psychological Medicine*, 45 (2015): 1061–1072.

133 C. Davenport, *Heredity in Relation to Eugenics* (New York: Henry Holt and Company, 1911).

134 G. Sweeney, *Fighting for the Good Cause: Reflections on Francis Galton's Legacy to American Hereditarian Psychology* (Philadelphia: American Philosophical Society, 2001), 102.

135 K. Malik, *The Meaning of Race*, *op. cit.*, 114.

136 D. J. Kevles, *In the Name of Eugenics: Genetics and the Uses of Human Heredity* (Berkeley, CA: University of California Press, 1985), 112.

Roosevelt (1858–1919), President Woodrow Wilson (1856–1924),[137] inventor Alexander Graham Bell (1847–1922),[138] president of Yale Irving Fisher (1867–1947) who was also president of the "Eugenics Research Association," and Supreme Court Justice Oliver Wendell Holmes (1841–1935). It was Holmes' ruling, in the case of *Buck v. Bell* in 1927, which allowed American states to institute eugenic policies of forced sterilization.[139]

In 1907, the state of Indiana passed a law to "prevent the procreating of confirmed criminals, idiots, imbeciles and rapists." By 1913, twelve other states had passed similar laws. However, in a test case in 1924, a superintendent of the "Virginia State Colony for Epileptics and the Feeble-Minded" filed a petition to sterilize one Carrie Buck (1906–1983). Carrie had a mental age of nine and her 52-year-old mother had a mental age of 8. Carrie's mother also had a record of prostitution, and three illegitimate children of whose paternity she was not sure. Carrie herself had an illegitimate child. In 1927, when the case was taken to the Supreme Court, the justices ruled that Carrie should be sterilized to protect the people of Virginia. Justice Oliver Wendell Holmes asserted:

> We have seen more than once that the public welfare may call upon the best citizens for their lives. It would be strange if it could not call upon those who already sap the strength of the State for these lesser sacrifices, often not felt to be such by those concerned, to prevent our being swamped with incompetence. It is better for all the world, if instead of waiting to execute degenerate offspring for crime, or to let them starve for their imbecility, society can prevent those who are manifestly unfit from continuing their kind. The principle that sustains compulsory vaccination is broad

137 A. Napolitano, *Theodore and Woodrow: How Two American Presidents Destroyed Constitutional Freedom* (Nashville, TN: Thomas Nelson, 2012).

138 N. Comfort, *The Science of Human Perfection: How Genes Became the Heart of American Medicine* (New Haven, CT: Yale University Press, 2012), 36.

139 Lynn, *Eugenics, op. cit.*, 27.

enough to cover cutting the Fallopian tubes.

Holmes added, citing an earlier case in another state, that: "Three generations of imbeciles are enough."[140]

As stated, the case was brought by the Virginia State Colony for Epileptics and the Feeble-Minded. Until successful treatments were developed, epileptics were regarded as a serious social problem. Due to their unpredictable and sometimes violent seizures, they were frequently unable to work or live independently. They were often institutionalized, initially alongside the "feeble-minded" and later in their own "colonies."[141] In a number of US states, until 1956, epileptics could be sterilised by court order and were banned from marrying. Until the 1970s, they could be legally excluded from restaurants, theatres, and cinemas.[142]

Between 1890 and 1929, 29 states banned the seriously mentally ill or the retarded from marrying and 19 states required couples to prove they were free of sexually transmitted diseases before they could marry.[143] By 1931, 30 US states had compulsory sterilization laws and, up until 1970, 60,000 Americans were sterilized under these laws, half of them being mentally retarded. The 1924 Immigration Act was partly motivated by evidence, from administering IQ tests to new immigrants, that there were very large numbers of "feeble-minded" people entering the United States from specific countries.[144] In California in 1934, a court-ruled that a 20-year-old heiress, Ann Cooper-Hewitt (1914–1956), could be sterilised—robbing her of her inheritance as her father had insisted that if she remained childless it would

140 A. Cohen, *Imbeciles: The Supreme Court, American Eugenics, and the Sterilization of Carrie Buck* (New York: Penguin, 2016).

141 D. Durst, *Eugenics and Protestant Social Reform: Hereditary Science and Religion in America, 1860–1940* (Eugene, OR: Wipf and Stock Publishers, 2017), Ch. 6.

142 S. Shorvon, "Neurology Worldwide: The Epidemiology and Burden of Neurological Disease" in C. Clarke, R. Howard, M. Rossor, and S. Shorvon, (Eds.), *Neurology: A Queen Square Textbook* (Chichester: John Wiley & Sons, 2016), 9.

143 Levine, *Eugenics, op. cit.*, 47.

144 J. Richards and R. S. Hawley, *The Human Genome: A User's Guide* (New York: Academic Press, 2011), Ch. 15.

revert to her mother—on the contested grounds of her being "feeble-minded." Eugenics author Paul Popenoe (1888–1979), the architect of the state's eugenic sterilisation program, argued that Ann's "sexual forwardness" was evidence of feeble-mindedness, and this was accepted. As a consequence, sterilisation was effectively legal in California on the grounds of promiscuous or non-conforming sexual behaviour.[145] In Britain, it was also common in this period for sexual promiscuity to be used as evidence that a woman was feeble-minded.[146]

The eugenics campaign even led to passive euthanasia. In 1915, Harry J. Haiselden (1870–1919), a Chicago surgeon, stunned the nation by revealing that he had allowed the deaths, with parental consent, of "at least six infants" he diagnosed as "defectives" by refusing to perform surgery on them. Haiselden "displayed the dying infants to journalists, wrote about them for the [...] newspapers, and starred in a feature film about his crusade." Prominent Americans including lawyer Clarence Darrow (1857–1938), defence counsel at the Scopes Monkey Trial, and disability-rights campaigner Helen Keller (1880–1968) "rallied to his support."[147]

A 1916 film called *The Black Stork,* written by Haiselden and a journalist, followed a storyline based on the case. Morally didactic, and clearly encouraging viewers with any "taint of the blood" not to have children, it was re-released in cinemas in 1918 as *Are You Fit to Marry?* It continued to be shown in small cinemas until 1942.[148] In keeping with support for eugenics among many religious people, Haiselden has the character based on and played by himself say, "I believe it wrong to operate to prolong an absolutely useless life" and, when challenged that it is his duty to save life, replies: "It is the will of God that this baby be born a defective and without the meddling of surgery it is the will of God that this child die. Shall I set myself up as wiser than the Almighty? God does not want this child to live." Unsurprisingly, considering the

145 Levine, *Eugenics, op. cit.*, 80–81.

146 P. Bartley, *Prostitution: Prevention and Reform in England, 1860–1914* (London: Routledge, 2012).

147 Pernick, *The Black Stork, op. cit.*, Blurb.

148 Pernick, *The Black Stork, op. cit.*, 6.

moral righteousness of some eugenicists and their influence on society, there were many anti-eugenics comedies and dramas at cinemas during this period as well. These films also tended to dispute hereditarianism more generally.[149]

Canada, Australia, and South Africa

Canada only practiced compulsory sterilisation of the "feeble-minded" in the provinces of Alberta (as of 1928) and British Columbia (as of 1933). These acts were repealed in the 1970s. There is also evidence of the coercive sterilisation of indigenous women across this period, such that by 1972, in Alberta, they were 25 per cent of all women who were compulsorily sterilised. Other provinces did not introduce such laws, partly due to resistance from Roman Catholics.[150] In 1912, eugenics was condemned in a study for the Catholic Social Guild by English priest Father Thomas Gerrard (1871–1916) for assuming humanity was animal in nature rather than spiritual. Eugenics, either positive or negative, was anathematized by Pope Pius XI (1857–1939) in an encyclical on 31st December 1930.[151]

Compulsory sterilisation laws were not introduced in the other main Anglophone countries either: Australia, where early twentieth century Prime Minister Alfred Deakin (1856–1919) was a strong supporter of eugenics,[152] New Zealand and South Africa. Instead, efforts were made to try to limit the fertility of poor whites and native peoples by strongly encouraging the use of contraception.[153] For example, free contraception was offered to poor Afrikaners in an attempt to limit what the British regard-

149 Pernick, *The Black Stork, op. cit.*, 130–131.

150 T. Marshall and G. Robinson, "Eugenics in Canada," *The Canadian Encyclopedia* (2006), https://www.thecanadianencyclopedia.ca/en/article/eugenics.

151 D. Kevles, *In the Name of Eugenics, op. cit.*, 127.

152 Levine, *Eugenics, op. cit.*, 12

153 See, D. Paul, J. Stenhouse, and H. Spencer, (Eds.), *Eugenics at the Edges of Empire: New Zealand, Australia, Canada and South Africa* (London: Palgrave Macmillan, 2017).

ed as their worryingly high fertility.[154]

Eugenics and Race in the US and the UK

Eugenics advocates, such as the American lawyer Madison Grant (1865–1937), who discussed the issue in his book *The Passing of the Great Race*,[155] were particularly concerned that immigration was destroying the Protestant Anglo-Saxon majority which was, Madison averred, biologically and intellectually superior in particular to Eastern and Southern European immigrants.[156] Fellow American Charles Goethe (1875–1966), sometime president of the Eugenics Research Association, was specifically concerned about Mexican immigrants interbreeding with White Americans and the impact that this would have on the nation's genetic quality.[157] Fears about the consequences of miscegenation led to the banning of inter-racial marriage in all but nine states between the seventeenth century and the 1910s. These laws were gradually overturned and when they were declared unconstitutional in 1967 sixteen states still banned interracial marriage, in most cases only between other races and blacks.[158]

Some British eugenicists also connected eugenics and race. For example, Scottish anthropologist Sir Arthur Keith (1866–1955), Rector of Aberdeen University, argued that it would ultimately be good for peace and stability on earth if less developed races simply perished. Fellow Scot, J. W. Gregory (1864–1932), Professor of Geology at Glasgow University, averred that "miscege-

154 S. Klausen, *Race, Maternity, and the Politics of Birth Control in South Africa, 1910–48* (London: Palgrave Macmillan, 2004). There was also a significant eugenics movement among British settlers in Kenya. See, C. Campbell, *Race and Empire: Eugenics in Colonial Kenya* (Manchester: Manchester University Press, 2013).

155 M. Grant, *The Passing of the Great Race: The Racial Basis of European History* (New York: Charles Scribner's Sons, 1916).

156 C. Rosen, *Preaching Eugenics: Religious Leaders and the American Eugenics Movement* (Oxford: Oxford University Press, 2004), 143.

157 S. Kuhl, *The Nazi Connection: Eugenics, American Racism, and German National Socialism* (Oxford: Oxford University Press, 2002).

158 Levine, *Eugenics, op. cit.*, 48; D. Harris and R. Morris, *Marriage, Race and the Law* (Minneapolis, MN: Abdo Publishing, 2020).

nation" should be stopped because it would dilute the superior qualities of successful peoples, such as the British.[159] He based this argument on observations from plants and other animals that "miscegenation" tends to produce "inferior progeny."[160] For Scottish anthropologist Sir James Frazer (1854–1941), Fellow of Trinity College, Cambridge and, briefly, Professor of Social Anthropology at Liverpool University, civilization could be seen to increase or diminish "with the blanching or darkening of the skin."[161] Moreover, the British lower classes were little different from the "savages" studied by anthropologists such as himself: "The truth seems to be that to this day the peasant remains a pagan and savage at heart; his civilization is merely a thin veneer which the hard knocks of life soon abrade, exposing the solid core of paganism and savagery below."[162]

Francis Galton espoused "the encouragement of the Chinese settlements at one or more suitable places on the East Coast of Africa as part of our national policy, in the belief that the Chinese immigrants would not only maintain their position, but that they would multiply and their descendants supplant the inferior Negro race" whom he described as "lazy, palavering savages" in contrast to the "order-loving Chinese." Galton further commented that the Chinese have a "remarkable aptitude for a high material civilization" but, alas, their culture "treats originality as a social crime."[163] Northeast Asians have a higher average IQ than whites but far lower levels of genius and originality, possibly due to a narrower IQ range, lower levels of psychopathology and lower levels of general curiosity.[164] As we will see in the next chapter,

159 G. Schaffer, *Racial Science and British Society, 1930–62* (London: Palgrave Macmillan, 2008), 46–47.

160 J. W. Gregory, *Race as a Political Factor* (London: Watts and Co., 1931), Abstract.

161 Quoted in: R. Ackerman, *J. G. Frazer: His Life and Work* (Cambridge: Cambridge University Press, 1987), 219.

162 J. Frazer, *The Golden Bough: A Study in Magic and Religion* (London: MacMillan and Co., 1919), ix.

163 F. Galton, "Africa for the Chinese," *The Times* (5th June 1873).

164 K. Kura, J. te Nijenhuis, and E. Dutton, "Why do Northeast Asians Win So Few Nobel Prizes?" *Comprehensive Psychology*, 4 (2015).

scientific genius tends to be characterized by outlier high IQ combined with moderately low Agreeableness and moderately low Conscientiousness.

Why Was the Nineteenth Century So Conservative?

Racial intermarriage became increasingly unacceptable from the middle of the nineteenth century onwards in tandem with the rise of the systematic study of racial differences or "race science."[165] This may have highlighted racial differences in a way that they had not previously been so systematically, so influencing attitudes. However, this increasing racial segregation, when taken together with increasing sexual conservatism across the nineteenth century,[166] may also have reflected broader trends. For example, it may have reflected a system of competitive conservatism-signalling; perhaps a conservative backlash in response to changes wrought by industrialization,[167] with heightened traditional religiosity tending to be a response to mortality salience, uncertainty and to feelings of exclusion.[168]

Increasing disease prevalence may have been another factor. As we have already explored, conservatism, which strongly crosses over with traditional religiosity,[169] is associated with disgust and especially with contamination disgust—concern with disease.[170] The Industrial Revolution led to large and fairly con-

165 D. van Galen Last, *Black Shame: African Soldiers in Europe, 1914–1922* (London: Bloomsbury, 2015), 14.

166 See, S. Garton, *Histories of Sexuality: Antiquity to Sexual Revolution* (London: Routledge, 2014), Ch. 6.

167 For a detailed discussion of political cycles see E. Dutton and J. O. A. Rayner-Hilles, *The Past is a Future Country: The Coming Conservative Demographic Revolution* (Exeter: Imprint Academic, 2022).

168 A. Norenzayan and A. Shariff, "The origin and evolution of religious pro-sociality," *Science*, 322 (2008): 58–62.

169 C. Lockhart, C. Sibley, and D. Osborne, "Religion makes—and unmakes—the status quo: religiosity and spirituality have opposing effects on conservatism via RWA and SDO," *Religion, Brain and Behaviour,* 10 (2020): 379–392.

170 Y. Inbar, D. Pizarro, R. Iyer, and J. Haidt, "Disgust Sensitivity, Political Conservatism, and Voting," *Social Psychological and Personality Science,*

stant movements of people (and thus to uncertainty and feelings of exclusion in new environments), to greater contact between unrelated people, to cramped and unhealthy living conditions, and thus to increased epidemics of previously relatively rare conditions that killed extremely quickly.[171] Thus, while industrialization and inoculation dealt with various illnesses, causing decreasing child mortality and a rising population overall, it concomitantly led to a dramatic rise in frightening illnesses that killed people very suddenly, most obviously cholera. This leads to vomiting, diarrhoea, dehydration, and death within hours of the onset of symptoms. There were also dramatic rises in typhoid and tuberculosis.[172] Venereal disease, especially syphilis, became a prominent public health issue towards the end the eighteenth century and continued to be throughout the nineteenth century, with it being increasingly understood that syphilis and other venereal diseases caused infertility and was massively present among soldiers and sailors, portending national defeat.[173] Incurable until the mid-twentieth century, venereal disease was a problem which could only be addressed via high levels of sexual control and taboo.[174] We would expect this increased "infection disgust" to substantially heighten conservatism and, conversely, as society was increasingly able to deal with epidemics, we would expect society to become less conservative. We would also expect the increased mortality salience involved in such diseases to increase religiosity.

Certainly, national-level parasite stress predicts national-level

3 (2011): 5.

171 S. Ellis and E. Ong, "Disseminated Infections: A Clinical Overview" in Y.-W. Tang, M. Sussman, D. Liu et al., (Ed.), *Molecular Medical Microbiology* (Amsterdam: Academic Press, 2015).

172 J. Mackenbach, *A History of Population Health* (Leiden: BRILL, 2020), Ch. 5.

173 J. Frith, "Syphilis: Its Early History and Treatment until Penicillin and the Debate on its Origins," *JMVH,* 20 (2021): 4.

174 L. Merians, "Introduction" in L. Merians, (Ed.), *The Secret Malady: Venereal Disease in Eighteenth-century Britain and France* (Lexington, KY: University of Kentucky Press, 1996); S. Cronberg, "The rise and fall of sexually transmitted diseases in Sweden," *Genitourinary Medicine,* 69 (1993): 184–186.

conservatism.[175] Research has found that a component of racial prejudice is the activation of a disgust response.[176] This makes sense because it would be adaptive to avoid strangers in case they carried novel pathogens.[177] Such elevated disgust could make sense of increased racial intolerance and segregation (as occurred in British India[178] and Singapore[179]), fear of and dehumanisation of the working class,[180] and also of the increased control of sexuality[181]: sexual promiscuity is more likely to lead to infection and the spreading of infection and it also evidences a lack of self-control which itself could spread infection.[182] Victorians understood that diseases were spread by contact or via some medium, such as water, but they didn't begin to understand precisely how—via microbes—until the end of the nineteenth century.[183] This hypothesis to explain why Victorians, and people in the nineteenth century more broadly, were so conservative in so many ways needs to be researched more, but it would seem to make sense of the Puritanical backlash which characterized the nineteenth century and much of the twentieth. It would have been set off by a combination of a conservative backlash against the Georgian Era, dramatic change, and feelings of exclusion, rising infection

175 J. Tybur, Y. Inbar, L. Aarøe et al., "Parasite stress and pathogen avoidance relate to distinct dimensions of political ideology across 30 nations," *PNAS*, 113 (2016): 12,408–12,413.

176 Y. Liu, W. Lin, P. Xu et al., "Neural basis of disgust perception in racial prejudice," *Human Brain Mapping*, 36 (2015): 12.

177 L. Aarøe, M. Osmundsen, and M. B. Petersen, "Distrust As a Disease Avoidance Strategy: Individual Differences in Disgust Sensitivity Regulate Generalized Social Trust," *Frontiers in Psychology*, (2016).

178 E. Agnew, *Imperial Women Writers in Victorian India: Representing Colonial Life, 1850–1910* (London: Palgrave Macmillan, 2017), 6.

179 R. Home, *Of Planting and Planning: The Making of British Colonial Cities* (London: Routledge, 2013), 125.

180 P. Ingham, *Language of Gender and Class: Transformation in the Victorian Novel* (London: Routledge, 2002), 109.

181 Garton, *Histories of Sexuality*, *op. cit.*

182 Tybur et al., "Parasite stress and pathogen avoidance relate to distinct dimensions of political ideology across 30 nations," *op. cit.*

183 F. Condrau and M. Worboys, "Epidemics and Infections in Nineteenth-Century Britain," *Social History of Medicine*, 22 (2009): 165–171.

disgust and rising sexual disgust, which then have created a kind of runaway conservatism, or runaway group-orientation, which continued even as medical breakthroughs were made and the pace of change slowed.

Eugenics in Continental Europe: Switzerland and Germany

Many countries in Continental Europe instituted negative eugenics laws. Switzerland was the first country to introduce such legislation in 1927, having been the first to pass eugenic marriage laws in 1907.[184] Until the early 1950s, recidivist rapists in Switzerland were given a stark choice: suffer castration and go free, or endure life imprisonment.[185]

In 1933, Germany passed the "Eugenic Sterilization Law" which allowed mandatory sterilization for sufferers from mental retardation, psychosis, or serious genetic defects. Between 1933 and 1939, approximately 400,000 Germans were sterilized under these laws and about half of those sterilized were mentally retarded.[186] More broadly, however, sterilisation was mandated for schizophrenia, "feeble-mindedness," Huntington's Disease, genetic blindness, genetic epilepsy, genetic deafness, severe physical deformity, and chronic alcoholism.[187] Nazi Germany, under the Nuremberg Laws of 1935, banned marriages between Germans and Jews, to preserve supposedly superior German qualities. It defined a Jew as a person with three Jewish grandparents, a German as one with four German grandparents and a "*Mischling*" as those in between.[188]

It also famously operated a campaign of what it regarded as

184 Levine, *Eugenics, op. cit.*, 47.

185 C. Krimbas, "Eugenics in Europe," in Neil Smelser and Paul Bates, (Eds.), *International Encyclopedia of the Social and Behavioral Sciences* (Amsterdam: Elsevier, 2001).

186 R. Proctor, *Racial Hygiene: Medicine Under the Nazis* (Cambridge, MA: Harvard University Press, 1988), 108.

187 A. Teicher, *Social Mendelism: Genetics and the Politics of Race in Germany, 1900–1948* (Cambridge: Cambridge University Press, 2020), Ch. 4.

188 A. Teicher, *Social Mendelism, op. cit.*, 199.

positive eugenics. Members of the SS were perceived to be an intellectual and certainly a physical elite. Thus, as part of Germany's *Lebensborn* program, acceptable "Aryan" women were matched with SS members to, in effect, breed more elite Germans. These people had to show pure German ancestry back to the mid-eighteenth century.[189] Approximately 20,000 children were born via this process. The two partners in any one example of this process never learnt each other's names and the female would be presented with a number of males, one of whom she would select. They would then sleep together until pregnancy was achieved. The resultant child would then be taken from the mother and raised in a special SS Home where it would be instilled with loyalty to Nazi ideals. After the War, many of these children were adopted and, with their birth records destroyed, were never able to discover their true origins.[190] Latvia also introduced a sterilisation law in 1933.[191] It has been argued, however, that, in many ways, the Nazis operated a clearly "dysgenic" policy. According to American psychologist Seymour Itzkoff, "[...] the Holocaust was a vast dysgenic program to rid Europe of highly intelligent challengers to the existing Christian domination by a numerically and politically minuscule minority."[192]

Nordic Countries

Sweden, a strongly socialist country from the 1930s onwards, sterilised 60,000 supposed undesirables between 1934, when its sterilization law was passed, and 1976. This was a much higher proportion of the population than had been sterilised in Nazi Germany.[193] Its strong welfare state—with accessible childcare

189 D. Hertz, "The Genealogy Bureaucracy in the Third Reich," *Jewish History,* 11 (1997): 53–78.

190 G. Milton, *Fascinating Footnotes From History* (London: John Murray, 2015), "The Women Who Gave Birth for Hitler."

191 P. McDonnell, *Disability and Society: Ideological and Historical Dimensions* (Dublin: Orpen Press, 2007).

192 S. Itzkoff, "Preface" in J. Glad, *The Future of Human Evolution: Genetics in the Twenty-First Century* (Schuylkill Haven, PA: Hermitage Publishers, 2008), 5.

193 Lynn, *Eugenics, op cit.*

and the right of women to work—was partly motivated by a desire to ensure that the more intelligent Swedish women would have more children.[194]

Neighbouring Finland instituted a similar law the following year, which was in operation up to 1970, though sterilization was far less frequent. It seems that coercive sterilization was more likely to be approved if the females were of low socioeconomic status and especially if they were part of Finland's Romani minority.[195] But, in theory, Finland's 1935 Sterilisation Law allowed for the compulsory sterilisation of "idiots," "imbeciles", and "the insane," defined as manic depressives and schizophrenics.[196] Finland's eugenics movement had actually been spear-headed at the beginning of the twentieth century by its small, but powerful and elite, Swedish-speaking minority, which substantially constituted its ruling class until around that time. This was partly motivated by a belief that the Swedes, who considered themselves Aryan, must rule the Finns, who were regarded as "Mongoloid," at least in part, and certainly as non-Aryan. The Finland-Swedes must, therefore, be more fertile than the Finns, eugenic activists reasoned.[197] Finland's eugenic law was also motivated by the cost of supporting the very poor.[198] A further law was passed in Finland in 1950 mandating compulsory abortion for the mentally retarded on eugenic grounds followed by the sterilization of the mother.[199]

194 Levine, *Eugenics, op. cit.*, 13.

195 M. Mattila, "Sterilization Policy and Gypsies in Finland," *Romani Studies*, 28 (2018): 109–139.

196 J. von dem Knesebeck, *The Roma Struggle for Compensation in Post-War Germany* (Hatfield: University of Hertfordshire Press, 2011), 36.

197 A. Spektorowski and L. Ireni-Saban, *Politics of Eugenics: Productionism, Population, and National Welfare* (London: Routledge, 2013), 80. For detailed discussion of the Finns see, E. Dutton, *The Silent Rape Epidemic: How the Finns Were Groomed to Love Their Abusers* (Oulu: Thomas Edward Press, 2019).

198 A. Elmgren, "The Nazis' Cloven Hoof: Finnish Critiques of Legal Sterilisation" in J. Kananen, S. Bergenheim, and M. Wessel, (Eds.), *Conceptualising Public Health: Historical and Contemporary Struggles over Key Concepts* (London: Routledge, 2018).

199 M. V. Seeman, "Sterilization of the Mentally Ill During the Years of

There were also eugenic sterilization laws in Denmark (as of 1929) and Norway (as of 1934) during this period.[200] For example, 3,800 people were compulsorily sterilized in Denmark between 1929 and 1945. Similar laws were passed in Iceland (1938) and Estonia (1936).[201] All of the Nordic countries had either Liberal- or Social Democrat-led governments when these laws were passed.

Catholic Europe and Latin America

Due to Catholic prohibitions on abortion, contraception, and sterilization, eugenics laws were not introduced in Catholic areas of Europe, though there were "positive eugenics" campaigns to increase fertility and to encourage people to contract judicious and healthy marriages.[202] Until papal condemnation of eugenics in 1930, there were many Catholic advocates for it, most notably German Catholic priest Hermann Muckermann (1877–1962) who, in the 1920s, was head of the Kaiser Wilhelm Institute for Anthropology, Human Heredity and Eugenics.[203] As of 1939, evidence of good health was required to marry in France.[204] In Franco's Spain, Antonio Vallejo-Nájera (1889–1960), Professor of Psychiatry at the University of Madrid, determined that the supposed social immorality of Marxists must mean that there were mentally ill and psychopathic,[205] possibly due to some kind

World War II in Finland," *International Journal of Mental Health*, 36 (2007): 58–66.

200 A. Drouard, "Concerning Eugenics in Scandinavia: An Evaluation of Recent Research and Publications," *Population*, 3 (1999): 261–270.

201 C. Krimbas, "Eugenics in Europe," in Neil Smelser and Paul Bates, (Eds.), *International Encyclopedia of the Social and Behavioral Sciences* (Amsterdam: Elsevier, 2001).

202 R. Cleminson, "Iberian eugenics? Cross-overs and contrasts between Spanish and Portuguese eugenics," 1930–1950, *Dynamics*, 37 (2017): 1.

203 H.-W. Schmuhl, *The Kaiser Wilhelm Institute for Anthropology, Human Heredity and Eugenics, 1927–1945: Crossing Boundaries* (New York: Springer, 2008), 105.

204 Levine, *Eugenics*, 47.

205 D. Pavon-Cuellar, *Marxism and Psychoanalysis: In or Against Psychology?* (London: Routledge, 2017, Trans. D. Kersey and D. Pavon-Cuel-

of "red gene."[206] Thus their children needed to be removed and raised by Fascist families for the children's own good, hoping to reduce the expression of the gene via environmental intervention. Between 1939 and 1950 approximately 30,000 children who it was feared carried the "red gene" were removed from Marxist parents by the Spanish government.[207]

There were eugenics movements in many Latin American countries, but no eugenic sterilisation laws were passed.[208] Mexico, however, passed a marriage law. Mexico's 1917 "Law of Family Relations" barred alcoholics, syphilitics and those diagnosed insane from marrying.[209] In the case of Brazil, as in other Latin American countries, this was partly because of a widespread belief in Lamarckian over Mendelian genetics, meaning that environment could influence genetics such that promoting sporting activities among the public was "eugenic."[210] However, a policy known as "whitening" was also espoused in Brazil, until around the time of World War I, as a means of dealing with the country's African slave-descended population. Whites should deliberately inter-breed with blacks to move the entire country closer to the ideals of "whiteness" and "civilization."[211] Other than very briefly under the Vichy Regime during World War II, there were never

lar), 1.

206 C. Pinhal, "Francoist Cinema: Orienting the Orient" in M. Grant, P. Ruzzante, and A. Hatton, (Eds.), *Iberian Studies: A (Trans-) National and (Trans-) Regional Exploration* (Newcastle upon Tyne: Cambridge Scholars Publishing, 2018), 109.

207 *El Pais,* "On the Trail of Spain's Stolen Children" (7th March 2011), https://english.elpais.com/elpais/2011/03/07/inenglish/1299478844_850210.html.

208 N. L. Stepan, *"The Hour of Eugenics": Race, Gender and Nation in Latin America* (Ithaca, NY: Cornell University Press, 1991).

209 Levine, *Eugenics, op. cit.*, 47.

210 N. L. Sepan, "Eugenics in Brazil, 1917–1940," in M. B. Adams, (Ed.), *The Wellborn Science: Eugenics in Germany, France, Brazil and Russia* (Oxford: Oxford University Press, 1990).

211 K. Caldwell, *Negras in Brazil: Re-envisioning Black Women, Citizenship, and the Politics of Identity* (New Brunswick, NJ: Rutgers University Press, 2007), 30.

negative eugenic laws in France either.[212]

Outside Europe

Positive eugenics laws—artificial insemination—and negative eugenics laws—sterilization—were proposed in the Soviet Union but were rejected.[213] Turkey passed eugenic marriage laws in 1930, by which time couples in Iran had to obtain a "certificate of wellness" before they could marry.[214] Japan passed eugenics legislation in 1949; overturning it in 1996. During this period, 16,520 Japanese females with undesirable characteristics were forcibly sterilised.[215]

India

Mahatma Gandhi (1869–1948), the spiritual founder of the Indian state who studied Law at University College London, was a convinced eugenicist. He wrote: "I should score out the condition of eugenic fitness because the begetting of offspring being the central purpose of marriage, eugenic fitness cannot be treated merely as a 'condition,' it is the *sine qua non* of marriage."[216] There were active branches of the Eugenics Society in various Indian cities by the 1920s.[217] India began a program of sterilization in 1975 and intensified this during its "State of Emergency" between 1976 and 1977 in which the Prime Minister, Indira Gandhi (1919–1984), ruled by decree. Sterilization was introduced ostensibly because India's population was felt to be too high. Poorer Indians were offered financial incentives to

212 D. B. Haycock, *Mortal Coil: A Short History of Living Longer* (New Haven, CT: Yale University Press, 2008), 193.

213 N. Krementsov, "Eugenics in Russia and the Soviet Union" in A. Bashford and P. Levine, (Eds.), *The Oxford Handbook of the History of Eugenics* (Oxford: Oxford University Press, 2012).

214 Levine, *Eugenics, op. cit.*, 47–48.

215 Lynn, *Eugenics, op cit.*

216 N. C. Beohar, *Wisdom of Mahatma Gandhi* (Chennai: Notion Press, 2018).

217 *Civil and Military Gazette*, "Indian Eugenics Society" (8th July 1921).

be sterilized. In 1976 alone, 7 million Indians were sterilized.[218]

China

China passed laws in 1945 permitting the compulsory sterilization of mental and physical defectives.[219] It has also been argued that Communist China's erstwhile "One Child Policy" was effectively eugenic, in that wealthier Chinese could afford to pay the fines for having more than one child and would thus be more likely to have more than one child than would poorer Chinese, with socioeconomic status being a robust correlate of intelligence and moral character as we have previously discussed. In the 1990s, a survey in China found that 82 per cent of doctors thought that a single woman on welfare with three children by three men should be sterilised and laws were introduced mandating compulsory abortion for women whose foetuses had serious genetic defects. Only 5 per cent of Western doctors who were asked the same question agreed.[220] Sterilisation was the most prevalent method of contraception in China in the year 2000.[221] Between 1979 and 1984 alone, 30 million women and 10 million men were sterilised in China.[222]

Singapore

In 1983 the Prime Minister of Singapore, Lee Kuan Yew (1923–2015), who ran the country from 1959 to 1990, was so concerned about dysgenics that he introduced tax rebates for graduate mothers who had children. Further, if they had three or more children then these children would be given priority access to

218 C. H. Brown, "The Forced Sterilization Program Under the Indian Emergency: Results in One Settlement," *Human Organization,* 43(1984): 49–54.

219 Y. J. Chung, *Struggle for National Survival: Eugenics in Sino-Japanese Contexts, 1896–1945* (New York: Routledge, 2002).

220 Lynn, *Eugenics, op cit.*

221 S. Short, M. Linmao, and Y. Wentao, "Birth planning and sterilisation in China," *Population Studies,* 54 (2000): 279–291.

222 J. Glad, *The Future of Human Evolution: Genetics in the Twenty-First Century* (Schuylkill Haven, PA: Hermitage Publishers, 2008), 50.

the most desirable schools. In his speech at the National Day Rally in 1983, Lee told his audience:

> If you don't include your women graduates in your breeding pool and leave them on the shelf, you would end up a more stupid society. So what happens? There will be less bright people to support dumb people in the next generation. That's a problem. If we continue to respond ourselves in this lopsided manner we will be unable to maintain our present standards. Levels of competence will decline. Our economy will falter, the administration will suffer, and society will decline [...] for every two college graduates in 25 years' time there will be one graduate and for every two uneducated workers there will be three.[223]

In 1984, Lee's government introduced grants to encourage poorly educated women to be sterilized and further incentives, such as priority access to the most desirable schools for poorly educated couples with at least three children who agreed to be sterilised. Due to a public uproar, most of these laws were abandoned or modified the following year.[224]

North Korea

Communist North Korea engages in a number of eugenic practices at the time of writing. According to defectors, those who are considered to be physically handicapped are not permitted to live in the cities. They are exiled to special institutions in the provinces and are not allowed to breed. Dwarves are arrested and sent to camps, where they are also prevented from having children.[225] North Korean female defectors who are returned suffer forced abortion and even the killing of their babies, supposedly because the children are assumed to be half-Chinese and thus

223 J. J. Palen, "Fertility and Eugenics: Singapore's Population Policies," *Population Research and Policy Review,* 5 (1986): 3–14.

224 J. J. Palen, "Fertility and Eugenics: Singapore's Population Policies," *op cit.*

225 H. Yahya, *Communism in Ambush: How the Scourge of the Twentieth Century is Preparing* (Istanbul: Global Publishing, 2003), 149–150.

impure.[226] North Korea also punishes the families of defectors or dissidents by sending several generations of the family, and even distant relatives, to prison camps.[227] This will, presumably, limit their fertility and, in so-doing, limit, in some small way, the partly-genetic traits that are associated with rebellion.

We have looked at the history of eugenic policies and the rise of the Eugenics movement. Evidently, it was an extremely influential movement in Western countries. As English historian Philippa Levine has summarised: "By the early years of the twentieth century eugenics had moved from paper to policy. Under the broad umbrella of eugenics, physicians and psychologists, social campaigners, feminists, and politicians of all stripes found common cause in a desire to use the findings of science to create a better world [...] In the first half of the century eugenics reached into many areas of life, from marriage and childrearing to criminality, from immigration to health care."[228] Clearly, this situation has now dramatically changed and in the next chapter we will explore how and why.

226 A. Lee, *North Korean Defectors in a New and Competitive Society: Issues and Challenges in Resettlement, Adjustment, and the Learning Process* (Lanham, MD: Lexington Books, 2015), 37.

227 E. Bradshaw, *North Korea: A Closer Look at the Secret State* (New York: Greenhaven, 2020), 83.

228 Levine, *Eugenics, op. cit.*, 2–3.

Nine

The Fall of Eugenics

Welfare Queens

Every so often, in the UK, a case is reported in the newspapers of a woman with a large number of children, usually an unemployed single-mother, complaining about the tax-payer funded welfare system upon which she is parasitic: her welfare benefits—sometimes higher than the annual salary of an average job due to "child benefit"—are not high enough, her tax-payer funded council house isn't big enough, and so on.[1]

In November 2021, 39-year-old Marie Buchan from Birmingham, an unemployed single mother of eight children, complained that she was forced to use "food banks" because two of her children were now over the age of sixteen, meaning she no longer received their child benefit payments. Having been on television to boast about milking the benefits system, and consequently becoming known as "octomom" and "welfare queen," Buchan had to endure some neighbours expressing their displeasure. "Some of the abuse I receive is just vile. They call me a parasite [...]"[2] A

1 In the UK, council houses are owned by local authorities and distributed to local tenants according to perceived need. For example, if a couple with a child could not afford a suitably sized private sector property they might be offered a council house. The rent for this is sometimes partly or wholly subsidized by the tax payer and there tends to be a waiting list for council houses. See, M. Harloe, *The People's Home? Social Rented Housing in Europe and America* (Hoboken, NJ: John Wiley & Sons, 2008).

2 J. Rose, "Jobless mother-of-eight says she is forced to use food banks

common reaction to such a person is something like: "If people can't afford to look after their children without outside help they have no business making copies of themselves."[3] But few would dare say this in public and under their own name.

We all know what the future consequences of the ubiquity of such women are likely to be in terms of national intelligence and moral character. But it's a taboo subject. We can't openly discuss it. In the 1920s, we *were* openly discussing it and we were trying to remedy it. In the UK, Buchan would probably have been diagnosed as "feeble-minded" and likely have had her children taken away and adopted. In parts of the US, she might even have found herself being sterilised. Why did the Eugenics Movement collapse?

The Wall Street Crash

Some commentators, such as Adam Rutherford, whom we met earlier, argue that it declined because the Nazis practiced it and its "horrors" came to light after World War II: "The toxicity of the idea of eugenics no doubt emerged from our collective discovery of the horrors of the Second World War."[4] This is simply historically inaccurate. We will see below that eugenics was advocated by many prominent, mainstream British academics until around the early 1960s.

Eugenics began to decline in popularity with the Wall Street Crash of October 1929 and the consequent Great Depression. As American literary scholar Daylanne English has argued, the Great Depression "was clearly an external force" and "many middle—and even upper-class individuals were adversely affected." It "forced many eugenics advocates to realize that unemployment did not necessarily signify 'individual weakness or a lack of abil-

to feed her children after benefits were cut by £160-a-week because her eldest two are now over 16," *Mail Online* (28th November 2021), https://www.dailymail.co.uk/news/article-10251865/Mother-eight-says-forced-use-food-banks-feed-children-benefits-cut.html.

3 Adam Perkins, Twitter (5th November 2022) with reference to a different case.

4 A. Rutherford, *Control: The Troubling History and Dark Present of Eugenics* (London: Orion, 2022).

ity to compete successfully in the marketplace.""[5] In other words the desire to preserve a positive self-image during an economic disaster put many previously wealthy people off of eugenics.

The 1960s: A Shift in Values

Then a fundamental shift in values took place in Western societies, especially in the 1960s. There was a shift in the dominant set of "moral foundations" that we explored earlier, from "binding" to "individualistic." As we discussed earlier, in Western countries, those who are labeled "conservative" score about the same on all five moral foundations. Those who are called "liberal" score high on the individualizing foundations but low on the binding foundations. But there are some nuances here. Each foundation is composed of sub-foundations. Thus, today's liberals score low in generalized "disgust" but very high in "moral disgust," which may explain why self-professed "liberals" cannot tolerate those who disagree with them on what they see as "moral" issues.[6] It has, though, been argued that this "moral disgust" appears due to measurement problems and that liberals are actually experiencing "moral anger."[7]

The key result of this liberal-conservative imbalance in moral foundations, however, is that the right can empathize with the left, but the left has more difficulty empathizing with the right. Consequently, the left tend to take over the machinery of culture and—up to a point—push it in a more left-wing direction, even if the population is becoming more right-wing for genetic reasons.

5 D. English, *Unnatural Selections: Eugenics in American Modernism and the Harlem Renaissance* (Durham, NC: University of North Carolina Press, 2005), 225.

6 See J. Graham, J. Haidt, and B. Nosek, "The Moral Stereotypes of Liberals and Conservatives: Exaggeration of Differences across the Political Spectrum," *PLOS One* (2012), https://doi.org/10.1371/journal.pone.0050092 and J. Graham, J. Haidt, and B. Nosek, "Liberals and Conservatives Rely on Different Sets of Moral Foundations," *Personality Processes and Individual Differences,* 96 (2009): 1029–1046.

7 M. Oaten, R. Stevenson, M. Williams et al., "Moral Violations and the Experience of Disgust and Anger," *Frontiers in Behavioural Neuroscience,* (2018).

The right, able to empathize with the left, will also cede ground to them. This is a general process. However, our own situation is unique in world history and requires a further level of explanation.

As already discussed, we were under harsh Darwinian selection to be group-oriented until at least 1800. Accordingly, with the collapse of harsh Darwinian selection, there would be a slow but steady accumulation of damaging mutations, and this would be associated with deviation from this "norm," with being an "individualist." Consistent with this, we have already seen that being left-wing correlates with low Agreeableness, low Conscientiousness, high mental instability, Machiavellianism and Narcissism. It is, in essence, being an individualist. We are evolved to be with other group-oriented people, and we are evolved to, in essence, cede ground to individualists. The result is that even relatively group-oriented people (though not those who were very strongly genetically group-oriented; they would be resistant) are increasingly inculcated with these particular kinds of leftist views, especially after a "tipping point" is reached, as it has been shown that once this happens there is mass movement towards what is regarded as the up-and-coming and soon to be triumphant perspective. Experiments have demonstrated that once a "critical mass" in a group—approximately 20 per cent—espouse a counter-cultural perspective, the rest of the group seems to very quickly adopt it.[8]

It appears that, in Western countries, this shift towards "individualizing" took place in the 1960s. At this point academics, journalists, politicians, and public intellectuals realized that it was better for their careers to espouse the new paradigm, to the point that they ended up believing in the new paradigm themselves via effortful control. The result, in the absence of much pressure against it, has been runaway individualism. The use of individualistic words in American texts increased markedly between 1960 and 2008.[9] The dominant set of values is individ-

8 D. Centola, J. Becker, D. Brackbill, and A. Baronchelli, "Experimental evidence for tipping points in social convention," *Science,* 360 (2018): 1116–1119.

9 J. Twenge, W. K. Campbell, and B. Gentile, "Increases in Individualistic

ualistic. Intelligent people better understand what the dominant set of values is, and the benefits of conforming, and have the effortful control necessary to persuade themselves of the veracity of the dominant set of values, allowing them to better attain status.[10] This creates an "arms race" where such people compete to signal their individualism, pushing society in an ever more left-wing direction. Thus, the "liberation" of women moves to "homosexuals" and then "transsexuals" and then, what next? This also leads to a situation in which, eventually, individualism becomes maladaptive. You signal that you are so moral, and white people are so evil, that you renounce having children, for example, and you encourage other white people to do likewise. Your ideas are "clever-silly" in the sense that they are espoused by highly intelligent people but are simply empirically incorrect or highly inconsistent, such as the idea that a person can "change gender."[11] As this becomes ever more extreme, we have the conditions of a new, restrictive religion in which nothing—certainly not truth, which in some ways is a matter of order and thus a binding value—is more important than "harm avoidance" and "equality." The result is the development of dogmas such as that "all people are equal in their innate capacities." This must be so, because to doubt it would upset people ("cause harm") and would even "promote inequality."

Indeed, in that intelligence is being eroded by dysgenics and this makes people low in trust and paranoid, the real-world dysfunction that this kind of belief system creates will be attributed not to the belief system, but to scapegoats such as "racists," "systemic racism," and "Russian hackers." Low intelligence will also mean sufficient deficiencies in reason such that people cannot understand why these views are illogical; as they will be governed

Words and Phrases in American Books, 1960–2008," *PLoS One*, 7 (2012); e40181.

10 M. A. Woodley of Menie and C. Dunkel, "Beyond the Cultural Mediation Hypothesis: A reply to Dutton (2013)," *Intelligence*, 49 (2015): 186–191.

11 E. Dutton and D. Van der Linden, "Who are the 'Clever Sillies'? The intelligence, personality, and motives of clever silly originators and those who follow them," *Intelligence*, 49 (2015): 57–65.

by emotion and cognitive biases, such as agency over-detection.[12]

"Not Even Trying"

This process would, of course, render "eugenics"—which is about the good of the group over the individual—highly taboo. In addition, our modern stress on "equality" also renders eugenics taboo, because eugenics assumes that some traits are superior to others, and this is abhorrent to those who purport to believe in "equality." This is despite the fact that, for many eugenicists, there were humanitarian ideals behind their thinking. Furthermore, there is a degree to which eugenics, no matter how humanitarian, is inherently coldly logical, something that is intrinsically appalling to people who claim to be focused on love and harm avoidance.

In Galton we already had the humanitarian element which argued, in essence, that being born with bad genes is a burden for the genetically defective individual who does most of the suffering, in addition to being bad for society. These early eugenicists for the most part did not pursue the good of the collective only, but primarily individual well-being. In that sense they were the leftists of their time. Whatever the aim of eugenics—be it the individual or the nation or some other collective—eugenics is always highly rationalist and technocratic, based on a consequentialist ethic. Humans have difficulty reconciling this with a humanitarian aim of preventing suffering, which in human psychology is based on emotion. The rationalism and consequentialism of positive eugenics is psychologically more compatible with seeking one's advantage, with striving for power, sex, or money. This is, after all, the purpose for which intelligence and rationality have evolved in our species. As we have discussed, leftists are strongly motivated by power and status. However, high in Neuroticism—and thus in feelings of fear—they use the covert tactic of virtue-signaling about individualist moral foundations in order to achieve it.[13] In a sense, the positive eugenicist's approach em-

12 See, P. Boyer, *Religion Explained: The Human Instincts That Fashion Gods, Spirits and Ancestors* (London: Heinnemann, 2001).

13 J. F. Benenson, "The development of human female competition: al-

bodies a part of the leftist which the leftist despises, so the leftist projects his self-loathing onto the positive eugenicist in order to over-compensate for these feelings of self-doubt. Projection essentially involves denying the existence of a negative aspect of oneself in order to maintain a positive self-image. When one notices this trait in other people, negative feelings – including of powerlessness – are triggered, as one realizes that one's identity is disingenuous. By morally attacking the person who triggers these feelings, a sense of empowerment and self-importance is restored. Further, you can tell yourself – in the moment of being triggered - that you fight such behavior; it is not true of you. It is, in this way, a kind of quasi-Narcissism.

In a way, the supposedly cold-hearted positive eugenicist is brave enough to openly engage in the kind of "tough-minded" thought processes which the leftists, being cowards, fear to. Even the negative eugenicist can seem harsh and, in a context of runaway individualism, in which we have competitively virtue-signaled to a point where it is morally good to have Down's Syndrome sufferers because they make society more "diverse," he is a manifestation of "tough" thinking and the insecurities of the Machiavellian can be projected onto him. Sally Phillips, whom we met earlier, has argued that "screening" for Down's Syndrome is "a diversity issue—threatening the continued existence of a unique strand of humanity."[14] One really must ask: will little children on hospital wards have to be screaming in pain and begging to be killed in a chorus of excruciating agony before leftists stop competitively virtue-signaling and admit that health is better than sickness?

It is even better for the relatives of the sick. Parents of severely disabled children are exceptionally likely to divorce due to the stresses involved,[15] and are at significantly elevated risk of suf-

lies and adversaries," *Philosophical Transactions of the Royal Society B*, 368 (2013): 2013007920130079.

14 C. Owen, "This is how people reacted to the debate around a new test for Down's syndrome in unborn children," *Wales Online* (6th October 2016), https://www.walesonline.co.uk/lifestyle/tv/how-people-reacted-debate-around-11988626.

15 D. Flynn, *Severe Emotional Disturbance in Children and Adoles-*

fering from depression.[16] One memoir, written by the mother of a severely disabled child, poignantly describes the feelings involved: "Being overwhelmed by suicidal thoughts is not uncommon among people with disabled kids. Frankly, I'm surprised more parents of special-needs kids don't abandon their families or kill themselves."[17] Certainly, runaway individualism has gone so far that society promotes sickness rather than health. This is particularly obvious in the relatively recent rise of the "fat acceptance" movement, the promotion of fat models in the media and attempts to make "fat-shaming" as socially unacceptable as "racism."[18] For many people, seeing "plus-size models" (when this term is a euphemism for "obese models") promoted in the media will provoke feelings of disgust and, thus, of dysphoria.[19] The same is true with regard to those who evoke feelings of disgust due to obvious health syndromes. Where once they were kept away from public view due to shame about "bad blood," or kept in institutions, they are now actively promoted in the name of "equality." For example, BBC presenter Adam Pearson (b. 1985) has a severely disfigured face due to suffering from the same genetic condition as the "Elephant Man" Joseph Merrick (1862–1890).[20] BBC children's channel *CBeebies* has produced a program called "Something Special" in which a clown called Mr. Tumble spends time with mentally retarded children, many of whom have Down's Syndrome and other serious genetic abnor-

cents: Psychotherapy in Applied Contexts (London: Routledge, 2013), 168.

16 S. Harris, "The Family Crisis: The Diagnosis of a Severely Disabled Child" in M. Ferrari and M. Sussman, (Eds.), *Childhood Disability and Family Systems* (London: Routledge, 1987).

17 A Bristowe, *My Own Blood: A Memoir of Madness and Special Needs Parenting* (Toronto: Random House of Canada, 2021), 395.

18 S. Bograd, B. Chen, and R. Kavuluru, "Tracking sentiments toward fat acceptance over a decade on Twitter," *Health Infomatics Journal,* (2022); R. Donahue, "This Model Shut DOWN the Man Who Body-Shamed Her On a Flight," *Allure* (2nd July 2017).

19 L. Vartanian, "Disgust and perceived control in attitudes toward obese people," *International Journal of Obesity,* 38 (2010); 1302–1307.

20 *BBC News,* "You are the modern day Elephant Man" (19th January 2016).

malities.[21]

This emphasis on "equality" would mean that the truth would be suppressed, and that people would be promoted to positions of influence—such as being in charge of groups of scientists—not because they were intellectually the best, but because they ticked certain equality boxes. In a sense, we are "Not Even Trying" to be the best once we reach this stage.[22] Leftists are placing "feelings" over "truth." This is a problem whenever there is a conflict between telling the truth and making people feel comfortable. Typical males, and autistics and geniuses in particular, will prefer truth-telling, while women typically prefer not to make people feel uncomfortable by telling them upsetting truths. It should be obvious which of these is more conducive to the progress of science. One interesting argument is that empathy-driven concealment of true facts and promotion of falsehoods is likely to lead to good outcomes only when people are sufficiently knowledgeable and intelligent to correctly predict the consequences of the untruths that they disseminate. Take as an example the old argument that religion is nonsense, but should be promoted because it supports a benign value system that promotes social harmony. This empathically motivated strategy probably worked well because parts of the Christian value system are indeed superior. But claiming that all racial inequalities come from systemic racism has pernicious consequences because it sows discord and prevents constructive efforts to solve the problem. In a world where intelligence is declining, people can support such a strategy because they are both empathically motivated and unable to predict the consequences of the strategy.

Feminism and the Suppression of Genius

Feminism makes this decline worse still by suppressing genius. The genius combines outlier high IQ with moderately low Agree-

21 E. Martin, "CBeebies 'Mr Tumble' trends on Twitter and people are scared to look," *Wales Online* (26th April 2022), https://www.walesonline.co.uk/lifestyle/tv/cbeebies-mr-tumble-trends-twitter-23786578.

22 See, B. G. Charlton, *Not Even Trying: The Corruption of Real Science* (Buckingham: University of Buckingham Press, 2012).

ableness and moderately low Conscientiousness. This means that, being low in rule-following, the genius can "think outside the box" allowing him to make original connections which, when taken together with very high intelligence, permit one to generate brilliant innovations. However, new ideas always offend against vested interests; they "rock the boat." But, being low in Agreeableness, the genius either doesn't care about that or is too low in empathy to anticipate that his ideas would cause offence even if he did care. It may be that these traits stay in the population—manifesting every so often in geniuses—because the genius is "group-selected": the group with an optimum small number of geniuses, who make brilliant innovations, will triumph over other groups, all else being equal. The downside of these geniuses is criminals and dreamers, when these traits manifest with insufficiently high IQ or in the wrong balance, but this will be outweighed by the positive contribution of the geniuses.[23]

Feminism is a problem for the genius because it helps to drive geniuses out of the "safe spaces" that they have held for a period of time: universities. Aspects of autism are best explained as the expression of an "extreme male brain." Autistics are very high in systematizing and very low in empathy—rather like the prototypical genius. The extreme female brain is very high in empathy, but very low in systematizing: the opposite of the genius.[24] Indeed, women in general are higher in Agreeableness and Conscientiousness than men, and are also more conformist,[25] which may additionally be influenced by their higher social anxiety.

Thus, once females are allowed into the universities, we would expect them to take over, because their intelligence is little different from that of males (except there are more males with outlier low and outlier high IQ and thus male more geniuses),[26] and they

23 See, E. Dutton, *Race Differences in Ethnocentrism* (London: Arktos, 2019).

24 S. Baron-Cohen, "The extreme male brain theory of autism," *Trends in Cognitive Sciences* 6 (2002): 248–254.

25 A. Eagly and C. Chrvala, "Sex Differences in Conformity: Status and Gender Role Interpretations," *Psychology of Women Quarterly,* 10 (1986): 203–220.

26 E. Dutton, "The male brain, testosterone and sex differences in pro-

are higher in Conscientiousness, which is vitally important to doing well in the world of education.[27] Taking over, they would push departments, and eventually entire universities, towards the individualizing values of "equality" and "harm avoidance" which are important to females. These "Head Girl" types are exactly the opposite of the "genius." If they take over, which they have done, then the result is speech codes, the suppression of "offensive" research—with highly original research always offending vested interests—even if it is empirically accurate, and, in general, the suppression of the genius in favour of ensuring that everyone feels equally valued, safe, and loved.

The result is illogical, verbose pseudo-argumentation which can't be rigorously critiqued lest it hurt the person's feelings. The result is hearing everyone's "perspective" and regarding them as equally "valid" because they are of equal worth. Thus, the genius no longer has a "safe space" in which to operate, and universities revert to being what they were in the mid-nineteenth century—means of promoting the state religion; though, at that time, the state religion was group-selected rather than individualistic. This feminizing, anti-systematizing shift—in which institutions become giant nursery schools—has also occurred at the societal level. An analysis of texts between 1850 and 2015 found that we reached a peak of "rational" phrases (such as "analysis") in 1980, after which "rational" words declined and intuitive or "feelings" words (such as "believe") rose, and skyrocketed since the year 2000.[28]

fessional achievement," *Mankind Quarterly*, 58 (2017): 93–100.

27 M. Almlund, A. Duckworth, J. Heckman, and T. Kautz, "Personality, psychology and economics," in S. Hanushek, S. Machin, and L. Woesmann, (Eds.), *Handbook of the Economics of Education* (Amsterdam: Elsevier, 2011).

28 M. Scheffer, I. van de Leemput, E. Weinans, and J. Bollen, "The rise and fall of rationality in language," *Proceedings of the National Academy of Sciences*, 118 (2021): e2107848118

Casting a Large Net

By around the 1970s, the entire issue of "eugenics" was increasingly taboo, and the term "eugenicist" gradually morphed into a damaging and emotive insult, used to shut down unpalatable scientific debate, rather akin to "racist," as discussed earlier. In a sense, "eugenicist" simply meant "heretic" against the new "religion of individualism." Eugenics also suffered from a superficial association with Nazi Germany's overt eugenic policies, such as mandatory sterilisation of perceived defectives. It was, therefore, tarred by association with a regime that was widely regarded as wicked.

Academic journals of eugenics began changing their names: *Eugenics Quarterly* became *Social Biology* and the *Eugenics Review* re-launched itself as the *Journal of Biosocial Science.* As already discussed, the "Eugenics Society" changed its name to the "Galton Institute" and, considering Galton's less-than-woke views, one wonders how soon it will be before it changes its name again. Eugenics itself was increasingly dismissed as "pseudo-science" and as "white supremacist," especially in light of an increasingly popular, and manifestly empirically inaccurate, dogma that all psychological differences between groups and individuals could be explained in terms of environmental factors.

Researchers and politicians who dared to touch on issues such as dysgenics increasingly found themselves subject to outrage mobs and duly ostracised. The "mobbing" of academics whose research has led to accusations of racism has been explored in detail elsewhere[29] and should be separated from those who have specifically provoked outrage due to perceived advocacy of eugenics. As already explored, the net of the new academic insult of "eugenicist" is cast very wide. For example, it has been argued that if you are interested in race differences in intelligence, then this knowledge can only be useful to you if you wish to stop certain races from breeding, breeding with other "superior" races, or entering your country and so reducing its average intelligence.

29 N. Carl and M. A. Woodley of Menie, "A scientometric analysis of controversies in the field of intelligence research," *Intelligence,* 77 (2019): 101397.

You cannot possibly be interested in these issues simply to make sense of the world. Therefore, you are a "eugenicist." English sociologist Noah Carl (b. 1990) was dismissed by Cambridge University in 2019 in part for his study defending research into race differences,[30] and for his journal article on racial differences in criminality and on the empirical accuracy of racial stereotypes. Carl was labelled as a "eugenicist,"[31] despite his never having explicitly espoused this. In 2020, American physicist Steve Hsu (b. 1966) was forced to resign as senior vice president of Michigan State University for having stated that his genomics company might be able to spot embryos with genes that make high IQ more likely, and that people with high IQ are more useful to society than are people with low IQ. This made him a "eugenicist" as well as a "scientific racist" according to the online woke mob who petitioned for Hsu's removal.[32]

If you are not directly accused of being a "eugenicist" then you are effectively criticised for being "eugenics-adjacent": you read "eugenicists" without becoming enraged, you listen calmly to their arguments, you cite them with regard to other issues, or you simply have some of the same interests, such as the heritability, or otherwise, of psychological traits such as intelligence. In 1994, American researchers Richard Herrnstein (1930–1994) and Charles Murray published *The Bell Curve.* In this book they showed evidence that there were race and social class differences in intelligence, and also social class differences in fertility, which they demonstrated is negatively associated with socioeconomic status, and also that intelligence, which predicts socioeconomic status, is strongly genetic and is negatively associated with fertility. Citing English psychologist Raymond Cattell's (1905–1998)

30 N. Carl, "How Stifling Debate Around Race, Genes and IQ Can Do Harm," *Evolutionary Psychological Science,* 4 (2018): 399–407.

31 B. Van der Merwe, "No, objecting to Cambridge's appointment of a eugenicist is not about free speech," *The New Statesman* (20th December 2018), https://www.newstatesman.com/politics/education/2018/12/no-objecting-cambridge-s-appointment-eugenicist-not-about-free-speech.

32 M. Cook, "More protests over eugenics, this time at Michigan State," *BioEdge* (20th June 2020), https://bioedge.org/enhancement/eugenics/more-protests-over-eugenics-this-time-at-michigan-state/.

1938 article (which we will discuss later) on what a society with declining IQ would be like,[33] they averred that the future would involve a civilizational declension.[34] Clearly, they were not espousing eugenics. However, they were criticized by American biologist Stephen J. Gould (1941–2002)—who later falsified scientific data in an attempt to disprove the fact that there are modal race differences in cranial capacity, consistent with race differences in brain size and thus in average intelligence[35]—because their work "echoed eugenic thoughts and practices."[36]

More recently, in 2016, *The Guardian* insinuated that psychologist Adam Perkins, of King's College London, believed in eugenics because in his book *The Welfare Trait* he proposed that there appears to be a partly genetic "employment resistant personality," [37] and noted that eugenics could breed this out, without saying that it should.[38] In 2019, *New York Times* columnist Bret Stephens was "accused of eugenics" simply for stating the objective fact that Ashkenazi Jews are the most intelligent of all ethnic groups.[39] Such accusations, of course, are related to the "Moralistic Fallacy;" the manifestly erroneous belief that because somebody states a fact then they must think it is a good thing—the fact-value conflation. In that leftists are high in Neuroticism,

33 R. Cattell, "Some Changes in Social Life in a Community with a Falling Intelligence Quotient," *British Journal of Psychology*, 28 (1938): 330–450.

34 R. Herrnstein and C. Murray, *The Bell Curve: Intelligence and Class Structure in American Life* (New York: Free Press, 1994).

35 J. Lewis, D. DeGusta, M. Meyer et al., "The Mismeasure of Science: Stephen Jay Gould versus Samuel George Morton on Skulls and Bias," *PLOS Biology*, 9 (2011): 7.

36 J. Maltby, L. Day and A. Macaskill, *Personality, Individual Differences and Intelligence* (London: Pearson, 2010), 370.

37 A. Perkins, *The Welfare Trait: How State Benefits Affect Personality* (London: Palgrave Macmillan, 2016).

38 D. Foster, "Adam Perkins: 'Welfare Dependency Can Be Bred Out,'" *The Guardian* (9th March 2016), https://www.theguardian.com/society/2016/mar/09/adam-perkins-welfare-dependency-can-be-bred-out.

39 E. Helmore, "New York Times columnist accused of eugenics over piece on Jewish intelligence," *The Guardian* (28th December 2019), https://www.theguardian.com/media/2019/dec/28/bret-stephens-new-york-times-jewish-intelligence-eugenics.

and thus paranoid, we can see how they might genuinely come to this conspicuously fallacious conclusion. Decreasing IQ, and the widening IQ range at universities due to increasing university-participation, may also be relevant to understanding the rise of illogical arguments, especially in academia. Given current dysgenic trends, we should expect this and many other fallacies like it to become increasingly prevalent in the public sphere, and concomitantly ever more reactive, aggressive institutional policies and legislation to police speech and academia.

It has been demonstrated that liberal democracy itself requires a value system that is based on trust and the reasoned assessment of the consequences of one's actions.[40] Today's experience shows that Western societies are becoming increasingly unable to maintain this moral foundation of democracy, and therefore democracy is predicted to be an early victim of IQ dysgenics and rising mutational load.

The Diehard Eugenicists

Some researchers, however, did not abandon a belief in eugenics, by which I mean an explicit belief in the importance of eugenics, rather than simply researching race differences in intelligence or anything else which leads ideologues to deploy the slur "eugenicist." Indeed, by about 1960 they were so concerned about the way in which people had lost interest in eugenics—with many geneticists trying to create a clear boundary between their work and eugenics—that they began to revive it. By the 1970s, however, as discussed above, eugenics was successfully stigmatized,[41] with the dogma that individual outcomes were primarily a matter of environment having been concomitantly promoted.[42] There was, in other words, the beginning of a "cultural revolution" involving

40 T. Vanhanen, *The Limits of Democratization: Climate, Intelligence, and Resource Distribution* (Augusta, GA: Washington Summit, 2009).

41 E. Ramsden, "Confronting the Stigma of Eugenics: Genetics, Demography and the Problems of Population," *Social Studies of Science,* 39 (2009): 853–884.

42 N. Rafter, *Creating Born Criminals* (Chicago, IL: University of Illinois Press, 1997), 89.

a strong shift away from heredity and towards environment, as leftist beliefs in inherent human equality and thus in environmental determinism began to noticeably influence science.[43]

Mankind Quarterly

In 1959, the "International Association for the Advancement of Ethology and Eugenics" (IAAEE) was founded in Baltimore by American zoologist Robert Kuttner (1927–1987). Anthony James MacGregor (1929–2019) (an Italian-American born "Antonio Gimigliano") of the University of California at Berkley was secretary, while American anthropologist Donald A. Swan (1935–1981), of the University of Mississippi, was treasurer. It was funded by the Pioneer Fund, a fund initially established by American political activist Wickcliffe Draper (1891–1972) to advance eugenic research.[44]

The organization's first executive committee included Scottish anthropologist Robert Gayre of Gayre (1907–1996), who was a Fellow of the Indian National Academy of Science and Professor at the University of Saugor (in Madhya Pradesh),[45] Canadian geneticist and Fellow of the Royal Society Reginald Ruggles Gates (1882–1962) and American psychologist Henry Garrett (1894–1973) who was a sometime president of the American Psychological Association and also Professor of Psychology at Columbia University in New York. Various other eminent scientists were also members. In 1962, the IAAEE set up an academic journal entitled *Mankind Quarterly.* This was initially based in Edinburgh, where it was edited by Robert Gayre of Gayre between 1962 and 1978. Although it occasionally carried articles advocating eugenics, it mainly focused on issues such as race, and also intelligence and other individual differences.[46] This rather shocked certain commentators, who felt that as "Fascists"

43 See U. Segerstråle, *Defenders of the Truth: The Battle for Science in the Sociobiology Debate and Beyond* (Oxford: Oxford University Press, 2000).

44 W. H. Tucker, *The Science and Politics of Racial Research* (Urbana, IL: University of Illinois Press, 1994).

45 *The Herald,* "Obituary: Robert Gayre" (17th February 1996).

46 Tucker, *The Science and Politics of Racial Research, op. cit.*

"misused" racial research it should surely be left alone and that to fail to do so was "to reopen Pandora's Box."[47]

In 1978, the editorship of *Mankind Quarterly* passed to Roger Pearson (b. 1927), an English anthropologist, trained at UCL, who lectured at the University of Southern Mississippi. The journal, therefore, migrated to Washington, DC.[48] Pearson has penned various books in which he has openly advocated positive eugenics, such as *Heredity and Humanity: Race, Eugenics and Modern Science.* In this book, he argues that civilization is based around intelligence, that some races are more intelligent than others, and that a positive eugenics program should be pursued among whites.[49] German biochemist Gerhard Meisenberg (b. 1953) took over the editorship in 2015, and I edited the journal in 2019. Since 2020, the nominal editor of *Mankind Quarterly* has been English psychologist Richard Lynn (b. 1930), formerly Professor of Psychology at the University of Ulster.

Richard Lynn

In his book *Eugenics: A Reassessment,*[50] Lynn also makes the case for positive eugenics, arguing that the light of civilization may well move to China because their one-child policy—combined with a female desire to sexually select for status (due to high status males being better able to invest in them and their offspring, leading to greater survival in pre-history)—will have raised Chinese IQ. This is because wealthier, more intelligent Chinese will have had more children and simply paid the fines for breaking the one child law, while less intelligent Chinese will have practiced abortion and infanticide to get a boy. The result is a huge sex imbalance, meaning that only the most intelligent males will likely pass on their genes. However, even so, intelligence is in decline in China and educated females have significantly fewer

47 R. Wolters, *Race and Education, 1954–2007* (London, MO: University of Missouri Press, 2008), 95.

48 Tucker, *The Science and Politics of Racial Research, op. cit.*

49 R. Pearson, *Heredity and Humanity: Race, Eugenics and Modern* Science (Washington, DC: Scott-Townsend, 1996).

50 R. Lynn, *Eugenics: A Reassessment* (Westport, CT: Praeger, 2001).

children than less educated ones (since the abolition of the one-child policy), with many of them remaining childless. In other words, Western patterns are being followed in China.[51] Indeed, in many developing countries there is now a negative association between education and fertility, especially among females.[52] Education is the best contraceptive. This may be so in part because the more intelligent have wider intellectual horizons, and therefore are less interested purely in family. For the less intelligent, especially for less intelligent females, there simply isn't much beyond the family that they understand well and that they find interesting and important.

In *Eugenics*, Lynn responds to various criticisms. He refutes the idea that a policy of eugenics would reduce creativity by selecting out schizophrenics, noting schizophrenia seems to inhibit rather than elevate creative achievement. He observes that though psychopathic traits do seem to be elevated among those who are highly creatively accomplished, this must be balanced with the negative dimensions to society of having a high number of psychopaths, with the chaos, and ultimately, the lack of creative achievement that this would bring about due to the greater need to concentrate on defending yourself subtracting from time spent on creative endeavour. He counters the argument that eugenics would reduce the number of per capita geniuses—who combine outline high IQ with moderately psychopathic traits—by noting that these tend to be born, by genetic chance, to parents of high intelligence and normal pro-social personality. Moreover, as we have discussed, per capita genius was at its height among a generation born around 1800 who were, thus, the survivors of harsh Darwinian conditions. He also notes the tremendous financial cost of dysgenics, with many developed countries spending 0.3 per cent of GNP on treating and caring for schizophrenics. Lynn counters the idea that disabled children can bring parents great inspiration and joy by noting that such parents experience high

51 M. Wang, J. Fuerst and J. Ren, "Evidence of dysgenic fertility in China," *Intelligence*, 57 (2016): 15–24.

52 G. Meisenberg and R. Lynn, "The Future of Human Intelligence." Unpublished Manuscript.

levels of stress and have significantly elevated levels of divorce.[53]

Lynn evidently has no qualms about being referred to as a eugenicist. In his memoirs, Lynn recalls that he gave a dinner for intelligence researchers at the Oxford and Cambridge Club in London in July 2011 for those who had contributed to a special issue of the journal *Personality and Individual Differences* that was focused on his work. Lynn's colleagues, led by Danish psychologist Helmuth Nyborg (b. 1937), presented him with a sword, a pair of horns and a plaque inscribed: "Tribute to Richard the Lionheart Lynn for his long-standing contributions to Eugenics and Psychometrics, given by devoted colleagues on 27th July 2011, at the celebratory dinner at the Oxford & Cambridge Club, London."[54]

Helmuth Nyborg

Danish psychologist and Olympian Helmuth Nyborg, Professor of Psychology at Aarhus University, made international headlines in 2003 when he publically argued that poor women should be paid to not have children for eugenic reasons.[55] In 2011, he referred to the phasing out of eugenics as "the greatest scientific fraud of the twentieth century," averring that eugenics needed to be made "normal science" once again.[56]

William Shockley

American physicist William Shockley (1910–1989), who had received the Nobel Prize for Physics in 1956, caused a media storm when he argued that people with an IQ of below 100 should be

53 Lynn, *Eugenics, op. cit.*

54 R. Lynn, *Memoirs of a Dissident Psychologist* (London: Ulster Institute for Social Research, 2020), 372.

55 M. Phillips, "A leading scientist says the poor should be paid NOT to have children. Have we learnt nothing from history?" *Mail Online* (2nd October 2003), https://www.dailymail.co.uk/columnists/article-229964/A-leading-scientist-says-poor-paid-NOT-children-Have-learnt-history.html.

56 H. Nyborg, "The Greatest Collective Scientific Fraud of the 20th Century: The Demolition of Differential Psychology and Eugenics," *Mankind Quarterly*, 51 (2011): 241–268.

paid to undergo voluntary sterilization, which he stated on television in 1974, by which time eugenics was stigmatized. Shockley successfully sued a journalist in 1981, because that journalist had implied that such views effectively rendered Shockley a Nazi sympathiser. However, Shockley was awarded only a dollar in damages as the jury felt that Shockley's reputation, as a eugenicist and a "racist" (for having stated in 1974 that blacks had lower average IQ than whites for mainly genetic reasons), was not worth very much.[57]

Raymond Cattell

Raymond Cattell, who ended his career at the University of Hawaii, had first expressed concern about dysgenics in a 1938 journal article entitled "Some Changes to Life in a Community with Falling Intelligent Quotient,"[58] to which we referred earlier. Cattell's shockingly prescient predictions included people spending longer at school (to try to compensate for slower learning), a fall in academic standards and thus grade inflation, a growing unemployable underclass, decreasing levels of major innovation, greater delinquency and a greater cultural focus on sex (perhaps because low IQ people are more instinctive and less interested in the abstract or maybe because low IQ is part of a suite of fast LHS traits at the group level), growing political extremism (look at the rise of the woke mobs), anti-democratic attitudes and general totalitarianism, reduced social cohesion, and increasing imitation of previous cultural achievements, as seen in the constant "remakes" on television.

In 1972, he returned to this issue with his book *Beyondism: A New Morality from Science*, in which he set out a Neo-Galtonian case for positive eugenics and the creation of a religion with a eugenic focus.[59] This did not cause much of a stir at the

57 J. Shurkin, *Broken Genius: The Rise and Fall of William Shockley, Creator of the Electronic Age* (London: Macmillan, 2006).

58 R. Cattell, "Some Changes in Social Life in a Community with a Falling Intelligence Quotient," *British Journal of Psychology*, 28 (1938): 330–450.

59 R. Cattell, *Beyondism: A New Morality from Science* (New York: Pergamon Press, 1972).

time, but the issue returned in 1997 when Cattell was chosen to receive the American Psychological Association's "Gold Medal Award for Lifetime Achievement in the Science of Psychology." Cattell flew from his home in Hawaii to Chicago to receive the award at the association's convention. Then, two days prior to the award ceremony, academic leftist-activist Barry Mehler, then a lecturer at Ferris State University in Michigan and founder of the "Institute for the Study of Academic Racism,"[60] launched a media campaign against the elderly Cattell, accusing him of being sympathetic to eugenics, which he had also done in a research article.[61] To avoid any further controversy, Cattell simply refused the award and withdrew his name from consideration. He died a few months later.[62]

David Lykken

In 1998, American geneticist David Lykken (1928–2006), Professor of Psychology and Psychiatry at the University of Minnesota, argued that levels of violent crime had significantly increased since the 1960s and needed to be reduced. This could be achieved by forcing potential parents to apply for a "parenting licence." He further averred that couples who were unmarried, unemployed, or suffering from psychiatric problems should automatically be denied such a licence. In that the genetic component of criminality is high and poverty and lack of a father elevate criminality, Lykken argued that "The licensure of parenthood is the only real solution to the problem of sociopathy and crime."[63]

60 S. Ibrahim, "Professor mocks students as 'vectors of disease' in profane tirade," *New York Post* (14th January 2022).

61 B. Mehler, "*Beyondism*: Raymond B. Cattell and the New Eugenics," *Genetica*, 99 (1997): 153–163.

62 A. Murphy Paul, *The Cult of Personality Testing: How Personality Tests Are Leading Us to Miseducate Our Children, Mismanage Our Companies, and Misunderstand Ourselves* (New York: Simon and Schuster, 2010), 195.

63 D. T. Lykken, "The Case for Parental Licensure" in T, Millon, E. Simonsen, M. Birket-Smith and R. Davis, (Eds.), *Psychopathy: Antisocial, Criminal, and Violent Behavior* (New York: The Guildford Press, 1998), 138.

Eugenicist Bioethicists

In 2004, Australian philosopher Nicholas Agar (b. 1965), Professor of Ethics at Waikato University in New Zealand, espoused what he called "Liberal Eugenics." In effect, this is the idea that parents should be able to use medical technology to alter the nature of their offspring, such that their offspring are closer to achieving what the parents would subjectively regard as a "good life." Agar writes: "Prospective parents may ask genetic engineers to introduce into their embryos combinations of genes that correspond with their particular conception of good life. Yet they will acknowledge the right of their fellow citizens to make completely different eugenic choices. No one will be forced to clone themselves or to genetically engineer their embryos."[64]

Some bioethicists, such as American Jonathan Anomaly, have even argued that it may be morally wrong *not* to use medical technology—such as genetic engineering and embryo selection—as failing to do so involves "rolling the genetic dice by having children without thinking about the traits they will have."[65] A number of ethicists, including Australian philosopher Peter Singer (b. 1946), have made the point, to which I will return below, that most people are, in reality, eugenicists: "everyone who considers pre-natal testing justifiable, or who thinks women should be free to weigh genetic information in the selection of a spouse or a sperm donor is a eugenicist."[66]

64 N. Agar, *Liberal Eugenics: In Defence of Human Enhancement* (Oxford: Blackwell, 2004), 6.

65 J. Anomaly, "Defending Eugenics," *Monash Bioethics Review,* 35 (2018): 24–35.

66 W. Veit, J. Anomaly, N. Agar, P. Singer, D. S. Fleischman, and F. Minerva, "Can "eugenics" be defended?" *Monash Bioethics Review*, 39: 60–67. *c.f.* R. Sear, "Demography & the rise, apparent fall and resurgence of eugenics," *Population Studies,* 75 (2021): 201–220.

Public Outcries

The taboo has also impacted politicians. In 1974, Conservative Shadow Home Secretary Sir Keith Joseph (1918–1994), who was a major influence on "Thatcherism," gave a speech in Birmingham in which he lamented the fact that girls with low educational attainment were producing "a third of all births. A high proportion of these births are a tragedy for the mother, the child, for us all." These births, argued Joseph, were a tragedy because they were likely to be "problem children, the future unmarried mothers, delinquents, denizens of our borstals, sub-normal educational establishments, prisons, hostels for drifters." Although some prominent people, such as the traditional morality campaigner Mary Whitehouse (1910–2001), supported Joseph, the media response was generally to paint Joseph as a eugenicist who wanted to stop low-IQ people from breeding, not least because of his words, "The balance of our human stock is threatened."[67] This furore contributed to Joseph withdrawing from the Conservative Party leadership contest and endorsing Margaret Thatcher (1925–2013).[68]

Social Democrat politician and economist Thilo Sarrazin (b. 1945) discussed the issue of dysgenics among the native German population, as well as the high fertility of the country's Muslim population and the supposed dysgenic consequences, in his 2010 bestseller the title of which translates as *Germany Abolishes Itself*.[69] However, the main reaction was accusations of racism.[70] There was a further media furore in 2010 when recently ennobled British Conservative former Shadow Chief Secretary to the Treasury, Howard Flight (b. 1948), expressed concern that changes to the welfare system would encourage poorer people to breed,

67 A. Denham and M. Garnett, *Keith Joseph* (London: Routledge, 2014), 266–267.

68 C. Moore, *Margaret Thatcher: From Grantham to the Falklands* (London: Penguin, 2013).

69 T. Sarrazin, *Deutschland schafft sich ab: Wie wir unser Land aufs Spiel setzen* (Berlin: Deutsche Verlags-Anstalt, 2010).

70 G. Hentges, "Between >Race< and >Class< [*sic.*]: Elite Racism in Contemporary Germany" in W. D. Hund, C. Koller and M. Zimmermann, (Eds.), *Racisms Made in Germany* (Munster: LIT Verlag, 2011).

with implied negative consequences for societal abilities: "We're going to have a system where the middle classes are discouraged from breeding because it's jolly expensive. But for those on benefits, there is every incentive. Well, that's not very sensible" he informed the London *Evening Standard.* Liberal Democrat MP Bob Russell (b. 1946) told *The Guardian* that Flight's "comments are offensive and unacceptable. They are not appropriate for the 21st century."[71] In other words, whether Flight's remarks were true or not, they went against a dogma of "equality" and thus they simply should not have been said.

In 2011, Martin Harty (1919–2017), a Republican member of the New Hampshire House of Representatives, who was 91 at the time and was first elected the previous year, was forced to resign for openly supporting eugenics. He told a woman who managed a community mental health program that "the world is too populated" and there are "too many defective people" such as "the mentally ill, the retarded, people with physical disabilities and drug addictions—the defective people society would be better off without." He further explained that nature has a way of "getting rid of stupid people," and "now we're saving everyone who gets born."[72]

The London Conference on Intelligence

One of the most significant "moral panics" over "eugenics" in academia occurred in 2018 with the "discovery" of the "London Conference on Intelligence" which had been taking place each year at University College London since 2015. It was organized by English psychologist James Thompson (b. 1948), who was

71 P. Walker, "Tory peer Howard Flight apologises over poor people 'breeding' comments," *The Guardian* (25th November 2010), https://www.theguardian.com/politics/2010/nov/25/howard-flight-rebuked-no-10-poor-people-breeding.

72 S. Schoenberg, "Lawmaker advocates eugenics," *Concord Monitor* (11th March 2011), https://www.concordmonitor.com/Archive/2011/03/999754836-999754836-1103-CM?page=0,0; M. Cook, "New Hampshire's eugenics legislator resigns," *BioEdge* (18th March 2011), https://bioedge.org/enhancement/eugenics/new-hampshires-eugenics-legislator-resigns/.

an honorary Senior Lecturer at UCL. The existence of the conference was publicised by English conservative journalist Toby Young (b. 1963), who had attended one of the conferences, in an article in the academic journal *Intelligence*. Young wrote of the 2016 meeting:

> I discovered just how cautious scholars in this field can be when I was invited to attend a two-day conference on intelligence at University College London by the academic and journalist James Thompson earlier this year. Attendees were only told the venue at the last minute—an anonymous antechamber at the end of a long corridor called 'Lecture Room 22'—and asked not to share the information with anyone else. One of those present, on discovering I was a journalist, pleaded with me not to write about the fact that he was there—he didn't want his colleagues to find out.
>
> It was like a meeting of Charter 77 in Václav Havel's flat in Prague in the 1970s.[73]

I was at this conference, and we knew what the venue would be long in advance, but I suppose we must forgive a journalist a little hyperbole.

Young's article was then reported by the London University student newspaper[74] and then by satirical magazine *Private Eye*,[75] before becoming national news. For having attended the conference, Young was forced to stand down from a government position, the aim of which, ironically, was to promote greater freedom of speech on British university campuses. The London Conference on Intelligence—many of whose attendees had published in *Mankind Quarterly*—was quickly described as a "secret eugenics conference,"[76] despite only one study presented over

73 T. Young, "Opinion: 2017 Constance Holden Memorial Address: Liberal Creationism," *Intelligence*, 66 (2018): 2–7.

74 B. Van der Merwe, "Exposed: London's Eugenics Conference and Its Neo-Nazi Links," *London Student* (10th January 2018).

75 *Private Eye*, "Toby Young Breeds Contempt . . ." (11th January 2018).

76 K. Rawlinson and R. Adams, "UCL to investigate eugenics conference secretly held on campus," *The Guardian* (11th January 2018), https://www.theguardian.com/education/2018/jan/10/ucl-to-investigate-secret-eugen-

three years being about eugenics and despite it being advertised.[77] Many of the studies were, however, about the genetics of intelligence and about race, including race differences in intelligence. UCL conducted a formal enquiry into how this had been allowed to occur, and pledged to "decolonize" itself. This was, of course, particularly embarrassing for UCL because, as discussed earlier, it had previously been the world centre of eugenics research. By 2020, it had started to rename buildings that were named after its most important scientists, such as Galton and Pearson.[78] The London Conference on Intelligence had always taken place in the "Pearson Building."

As a direct result of attending the conference, and of the consequent mini-moral panic about eugenics and race research that it provoked, a number of the academics were "cancelled."[79] Richard Lynn was stripped of his emeritus professorship in the Psychology Department at the University of Ulster (a department he had founded in 1972),[80] Noah Carl was fired from Cambridge University in part for attending the conference and for publishing alongside some of its more "controversial" academics,[81]

ics-conference-held-on-campus.

77 M. A. Woodley of Menie, E. Dutton, A. J. Figueredo et al., "Communicating intelligence research: Media misrepresentation, the Gould Effect and unexpected forces," *Intelligence*, 70 (2018): 84–87.

78 A. McKie, "UCL to begin process of renaming buildings named after eugenicists," *Times Higher Education* (28th February 2020).

79 This rise in "cancellation," where a person is effectively socially excluded because they are "mean," may well be a consequence of the feminisation of society and increasing female influence upon it. It is a "female" tactic, as we have discussed. Insomuch as it is a covert status-gaining technique that is mediated by female high anxiety about direct conflict, we might also expect left-wing males to employ it. American biologist Joyce Benensen has noted that men strongly desire to build large coalitions based on competence and toughness, leading them to overlook perceived "meanness." Females want a small number of "alloparents" whom they can be sure are "nice," and with whom they are strongly bonded. Thus, they behave very differently. See, J. Benenson, *Warriors and Worriers: The Survival of the Sexes* (Oxford: Oxford University Press, 2014).

80 Lynn, *Memoirs of a Dissident Psychologist, op. cit.*, 226–227.

81 Carl, "How Stifling Debate Around Race, Genes and IQ Can Do Harm," *op. cit.*

Gerhard Meisenberg was sacked from his Professorship of Biochemistry at Ross University in Dominica,[82] and James Thompson resigned from his honorary senior lectureship at UCL.[83] An analysis showed that the cancelling of academics—ranging from condemnation by their university to being fired—sky-rocketed from 2015 onwards and especially in the wake of the furore over the conference.[84]

In early 2020, Dominic Cummings (b. 1971), who was at that time the chief adviser to British Prime Minister Boris Johnson (b. 1964), was pursuing a policy of finding highly intelligent outsiders—"misfits and weirdoes" in Cummings' words—to work as government advisers. He appointed Andrew Sabisky (b. 1994), who had a degree in Classics from Kings College London. It soon came to light that Sabisky had presented at the London Conference on Intelligence,[85] that he was a race realist and that he was a "eugenicist" because he believed that benefits claimants should be encouraged to have fewer children. Under media pressure, Sabisky resigned from his post.[86]

Black Lives Matter and COVID-19

Then in June 2020, during the "Black Lives Matter" moral panic set off by the death of black criminal George Floyd (1971–2020)

82 A. Ellwanger, "An Open Letter on Campus Culture," *American Mind* (12th October 2020), https://americanmind.org/salvo/an-open-letter-on-campus-culture/.

83 A. Fazackerley, "UCL eugenics inquiry did not go far enough, committee say," *The Guardian* (28th February 2020), https://www.theguardian.com/education/2020/feb/28/ucl-eugenics-inquiry-did-not-go-far-enough-committee-say.

84 Woodley of Menie and Carl, "A scientometric analysis of controversies in the field of intelligence research," *op. cit.*

85 L. Middleton, "Boris Johnson aid said black people are mentally inferior," *The Metro* (17th February 2020), https://metro.co.uk/2020/02/17/boris-johnson-aide-said-black-people-mentally-inferior-12252270/.

86 R. Mason, "Boris Johnson adviser quits over race and eugenics controversy," *The Guardian* (17th February 2020), https://www.theguardian.com/politics/2020/feb/17/boris-johnson-adviser-quits-over-race-and-eugenics-writings.

in Minneapolis, Gonville and Caius College Cambridge announced that, under pressure from "Black Lives Matter" campaigners, they would remove a stained-glass window memorialising Sir Ronald Fisher, who had founded Cambridge University Eugenics Society. This had been erected in 1989, but, according to the college, Fisher's views were "not fully appreciated in 1989." Prior to the decision, woke activists had graffitied, "Eugenics is genocide—Fisher must fall" on its "Gate of Honour."[87]

In August 2020, former Labour leader Jeremy Corbyn (b. 1949) branded the government's earlier proposed policy of achieving "herd immunity" against the COVID-19 Pandemic as "eugenic," herd-immunity having been recommended in March by the Conservative government's chief scientific officer. "The last time I discussed herd immunity was when I worked on a pig farm 40 years ago. It was absurd that actually [you] would build up herd immunity by allowing people to die. And so, while the government was going into eugenic formulas and discussing all this stuff, they were not making adequate preparations," said Corbyn.[88]

87 M. Rusby, "Cambridge college to remove window commemorating eugenicist," *The Guardian*, (27th June 2020), https://www.theguardian.com/education/2020/jun/27/cambridge-gonville-caius-college-eugenicist-window-ronald-fisher.

88 K. Proctor, "Jeremy Corbyn: Government lectured me about herd immunity," *The Guardian* (19th August 2020), https://www.theguardian.com/politics/2020/aug/19/jeremy-corbyn-governments-coronavirus-policy-was-eugenic-herd-immunity.

Ten

The Future of Eugenics and Dysgenics

The End of Hospitals

We began this book with the tragic case of Nicky Ratcliffe, who died of a serious nut allergy at the age of 18. What happened to Nicky reached the newspapers because it was so rare. Many people have serious nut allergies, but the wonders of modern technology allow them to survive and, indeed, pass on the propensity to be highly allergic to their children. This does not augur well for the future.

As we have seen, we are becoming less intelligent for genetic reasons, and average intelligence has been shown to be central to every aspect of developing and maintaining civilization. As intelligence continues to decline, we eventually won't be able to do things we used to be able to do. We won't have the excess resources—or simply the problem-solving ability—to maintain a complex society, let alone an expensive health service. This is very serious because we have not just been selecting for low intelligence. We have also been allowing genetic mental and physical health to deteriorate across generations both of which are genetically related to low intelligence.

The Underclass and Medical Chaos

Of course, what this will eventually result in will be an underclass who are increasingly ill and about whom little can be done. Having poor impulse control, this will be rendered even worse by outbreaks of sexually transmitted diseases, which are already increasing,[1] drug addiction, also increasing among this demographic,[2] and even refusals to use vaccines (or even medicines) due to the association between low IQ and low trust and thus a propensity to accept outlandish conspiracy theories, rather than merely conspiracy theories for which there is sound evidence.[3] This is also already taking place, with low socioeconomic status being associated with empirically highly questionable "anti-vax" beliefs.[4] This would make high IQ conservatives, who would be high in disgust due to their high "sanctity," avoid such people and be utterly repelled by them. Such people would, after all, be vectors of disease. To make matters worse, the economic pressure that such a growing group of genetically sick people would place on health services, in a population of decreasing IQ and thus decreasing efficiency and decreasing economic resources to fund them, would be intolerable.

Doctors would be under growing pressure and would be increasingly dealing with violent, unreasonable, and highly contagious people. We might expect doctors to start abandoning the public health service for private practice, which they are already doing in Britain at the time of writing, with more than a third of General Practitioners planning to quit in the next five years.[5] People with the money are increasingly turning to private health

1 G. Anguzu, A. Flynn, J. Musaazi et al., "Relationship between Socioeconomic Status and Risk of Sexually Transmitted Infections in Uganda: Multilevel Analysis of a Nationally Representative Survey," *International Journal of STD and AIDS*, 30 (2019): 284–291.

2 J. Van Draanen, C. Tsang, S,. Mitra et al., "Socioeconomic marginalization and opioid-related overdose: A systematic review," *Drug and Alcohol Dependence*, 214 (2020): 108127.

3 J.-W. Van Prooijen, "Why education predicts decreased belief in conspiracy theories," *Applied Cognitive Psychology*, 31 (2017): 50–58.

4 T. Tomeny, C. Vargo and S. El-Toukhy, "Geographic and Demographic Correlates of Autism-Related Anti-Vaccine Beliefs on Twitter, 2009–15," *Social Science and Medicine*, 191 (2017): 168–175.

5 N. Bostock, "More than one in three GPs planned to quit within five years even before COVID-19," *GP* (10th June 2021), https://www.gponline.com/one-three-gps-planned-quit-within-five-years-even-covid-19/article/1718769.

as the NHS collapses, meaning NHS-users will be increasingly unreasonable and genetically sick people.[6] To make matters worse still, younger people—high in individualising values and low in group-orientation and thus low in a desire to serve the community[7]—decreasingly wish to even become doctors.[8] "Millennials" are increasingly quitting group-oriented professions in favour of self-realisation, a phenomenon being referred to as "The Great Resignation."[9]

Eventually, this growing underclass won't be treated, and there will be neither the money nor the desire to give them welfare. This will be so because among those who have sufficient intelligence, and are thus the tax base, the key factor predicting fertility is conservatism. Specifically, in the top quartile of IQ in the US, religiosity and conservatism either predict fertility or predict less sterility while atheism and liberalism are simply correlated with sterility.[10] Conservatives and religious people, in general, show weaker selection against intelligence than do liberals and among American Mormons there is positive selection for intelligence.[11] Similarly, in the German parliament—which is clearly composed

6 J. Kollewe, "Private hospitals profit from NHS waiting lists as people without insurance pay out," *The Guardian* (18th September 2021), https://www.theguardian.com/society/2021/sep/18/private-hospitals-profit-from-nhs-waiting-lists-as-people-without-insurance-pay-out.

7 J. Twenge, "The Evidence for Generation Me and Against Generation We," *Emerging Adulthood*, 1 (2013): 11–16.

8 *O2*, "Move over doctors and teachers, kids today are dreaming of becoming vloggers and animators" (4th December 2017), https://news.o2.co.uk/press-release/move-doctors-teachers-kids-today-dreaming-becoming-vloggers-animators/.

9 K. Ng, "The Great Resignation: Know When it's Time to Quit Your Job—and How to Do It Right," *The Independent* (27th August 2021), https://www.independent.co.uk/life-style/jobs-employees-how-to-quit-b1910055.html.

10 E. Dutton and J. O. A Rayner-Hilles, *The Past is a Future Country: The Coming Conservative Demographic Revolution* (Exeter: Imprint Academic, 2022).

11 E. Kirkegaard and E. Dutton, "Will Intelligent Latter-day Saints and Smart Conservatives Inherit the Earth? Differential Selection for Intelligence in the USA Based on Religiosity and Conservatism," *Evolutionary Psychological Science* (2022).

of that nation's elite—the fertility of legislators correlates with how right-wing their party is.[12] As a result of all this, we can expect to see a society marked by increasing chaos and disorder, as tends to occur in the Winter of Civilization.

Cycles of Civilization

I have explored in my book *At Our Wits' End,*[13] the large body of evidence that civilizations tend to go through cycles. They begin under harsh Darwinian conditions, including harsh levels of group selection. This selects for intelligence and martial values. They develop civilization, with its attendant easier lifestyle. This weakens Darwinian selection pressures and society shifts towards individualism. Under these conditions of low mortality salience, intelligent people also start questioning the religion which has previously promoted group-adaptive traits as God's will. This is because religiousness is elevated by mortality salience, which they will be lowest in, and because intelligence is associated with rationality, the "disenchantment of the world," and alienation from one's own emotions. When combined with affluence, rationality leads people to question the purpose of family life and reproduction in favour of more rewarding pursuits, and eventually to question the purpose of life itself.[14]

Intelligence invariably begets a sense of control over one's life, with the consequence that people start limiting their fertility in the absence of a strong religious belief that they should do otherwise or in the relative absence of mortality salience. As every development economist knows, education is the best contraceptive. This may be because educated, intelligent people are more efficient users of contraception, though it may also be

12 J. Sänger, G. Madison, D. Becker, H. Rindermann and E. Dutton, "Fertility and Ideology: Fewer Children and More Globalist? The Case of German Politicians from Left to Right." Unpublished Manuscript.

13 E. Dutton and M. A. Woodley of Menie, *At Our Wits' End: Why We're Becoming Less Intelligent and What It Means for the Future* (Exeter: Imprint Academic, 2018).

14 E. Dutton and D. Van der Linden, "Why is Intelligence Negatively Associated with Religiousness?" *Evolutionary Psychological Science,* 3 (2017): 392–403.

because intelligence is associated with environmental sensitivity; they are more sensitive to an "evolutionary mismatch" and this creates dysphoria and makes them act in a maladaptive fashion, as they require the adaptive "roadmap of life" to a greater extent than less intelligent people. Certainly, depression mediates not wanting children.[15] And, in advanced societies, the anti-natalism of the higher—more intelligent—groups is always observed, for reasons we have already explored.

But, whatever the causes, intelligence and religiosity—and that which it promotes, such as patriarchy and ethnocentrism in the case of traditional state religions[16]—falls, as does fertility and especially the fertility of the higher classes, and you see feminism (female lawyers and imams in Baghdad; Roman women leading armies and holding enormous political power),[17] multiculturalism and immigration, nihilism, sexual liberty, increasing tolerance of homosexuality, materialism, equality-fetish and a welfare state of sorts ("bread and circuses" in Rome), the worship of musicians and entertainers rather than explorers and soldiers, polarisation as the society is decreasingly united by a sense of its own eternal importance, a shift towards individualism and a loss of martial, group-oriented values which are replaced by a decadent desire for self-realisation or compassion, as fewer and fewer have ever been inculcated with a need to struggle. This same process was noted during the decline of Athens, Baghdad, and Rome.[18] Key turning points are obviously landmarks such as the first non-Roman Emperor of Rome, Trajan (53–117), who reigned from 98 to 117, and was from modern-day Spain, as was

15 P. Schönegger, "What's up with anti-natalists? An observational study on the relationship between dark triad personality traits and anti-natalist views." *Philosophical Psychology*, 35 (2022): 1.

16 Marginalized religious sects will often promote "equality" in order to attain power and, once in power and subject to group selection, their theology will tend to develop accordingly. A sense of marginalization has been shown to elevated religiousness. See, A. Norenzayan and A. Shariff, "The origin and evolution of religious pro-sociality," *Science*, 322 (2008): 58–62.

17 See, E. Dutton, *Witches, Feminism and the Fall of the West* (Whitefish, MT: Radix, 2021), 231–232.

18 J. Glubb, *The Fate of Empires and the Search for Survival* (Edinburgh: William Blackwood & Sons, 1976).

his successor, Hadrian (76–138).[19]

In this context, there is declining trust as people are no longer held together by the same religion, there is ethnic conflict, and there is falling trust even within ethnic groups, partly because more Machiavellian members collaborate with foreigners to attain status over their own group.[20] There is a gradual degeneration into war and polities start to split up. The more intelligent and group-oriented people manage to maintain civilization—as they did when Roman Civilization retreated to Byzantium—but when civilization finally collapses, harsh Darwinian conditions are re-imposed. The empire is duly displaced as the centre of civilization by another which is at an earlier place in the cycle, much as Rome displaced Athens.

Figure 3: Per Capita Major Innovations, 850 BC to 1950 AD

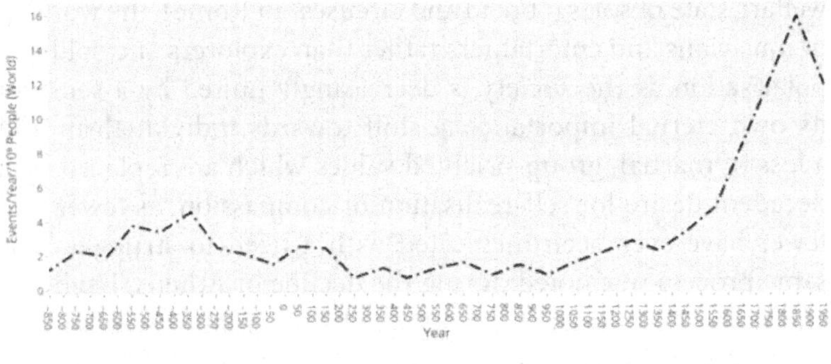

19 T. J. Dandelet, *The Renaissance of Empire in Early Modern Europe* (Cambridge: Cambridge University Press, 2014), 92. In modern times, we have the first non-British Prime Minister since the anomaly of the ethnically Jewish Benjamin Disraeli (1804–1881), Rishi Sunak (b. 1980). A man of Indian ethnicity, Sunak was appointed in 2022 in a cabinet with a number of South Asians and other ethnic minorities. This may occur partly because immigrants have been subject to harsher selection for longer, making them more socially effective and more religious, and partly because feelings of exclusion elevate religiosity and the related adaptive traits. Furthermore, immigration tends to be predicted by intelligence, and the ideology of Multiculturalism may lead to the over-promotion of immigrants.

20 R. Putnam, "*E Pluribus Unum*: Diversity and community in the twenty-first century. The 2006 Johan Skytte Prize lecture," *Scandinavian Political Studies*, 30 (2007): 137–174.

In the cases of Athens and Rome, we can see a peak of, and decline in, per capita major innovation consistent with the fall of civilization. We know that per capita major innovation more than doubled between 850 BC and 350 BC before going into decline, reaching a nadir in 50 BC, and only reaching the levels of 350 BC again in about 1600. Rome, which displaced Greece, reflected its own mini-cycle of innovation and intellectual sterility within a broader period of decline.[21] Not only that, but we know from a sample of Greek Ancient genomes that the prevalence of alleles associated with high intelligence was higher among Classical Greeks than among Modern Greeks, consistent with IQ decline.[22]

Mini-Cycles

It is worth, as a brief aside, exploring the concept of mini-cycles. A population can be broadly in decline, but undergo a mini-revival, or it can be broadly on the rise and undergo a mini-decline, meaning that the process is not exactly linear. For example, during the rise of civilization, a sudden cold period could lead to widespread war, resulting in a reduction in per capita major innovation and even in literacy, before the situation stabilized. Random events, such as a weak king, could have a similar impact. Conversely, during a decline, an important innovation might be made which might set off other important innovations, leading to what might be regarded as a "mini-revival," before the general process continued. In this regard, per capita major innovation fell considerably in both the mid-sixteenth and mid-seventeenth centuries, though the general movement was upwards. It also increased between about 1985 and the year 2000, due to innovations in computing, though the broad trend was downwards.[23]

Related to this, societies go through political and economic

21 J. Huebner, "Response by Jonathan Huebner," *Technological Forecasting and Social Change*, 72 (2005): 995–1000.

22 M. A. Woodley of Menie, "Cognitive Archeogenetics of Ancient and Modern Greeks," *London Conference on Intelligence* (May 2019).

23 J. Huebner, "A possible declining trend for worldwide innovation," *Technological Forecasting and Social Change*, 72 (2005): 980–986.

cycles. As we have seen, conservatives are about the same in all five moral foundations while liberals are only interested in individualizing foundations. This results in asymmetric empathy and a consequent tendency for conservatives to cede ground to liberals. This continues until dysphoria and chaos are created and there is, thus, a conservative backlash, such as took place in the UK in the 1979. This backlash may continue to the point of dysphoria, due to insufficient attention being paid to individualizing foundations, so there is a liberal backlash, as occurred in the UK in 1997.[24] We would predict that extreme liberalism, which tends to place "equality" over "truth," would harm innovation, but so might extreme conservatism, if it placed "loyalty to the group" or "religious dogma" over truth. In terms of economic cycles, there appear to be periods where people feel positive (leading to economic growth) and periods where people feel negative, resulting in recession. These may reflect natural adaptive cycles in how we feel across time—it being maladaptive to be too happy for too long—because anxiety incentivises planning and caution.[25] Innovation is easier to make if you don't have to worry about money. Thus, there would be certain optimum periods for intellectual innovation, and these could manifest even in an overall decline.

Population Collapse

But what will eventually occur if we continue selecting for the traits for which we are selecting? Under these conditions, with a people with very poor genetic health and very low intelligence, we would predict an unprecedented population collapse, as so many people would be reliant for their survival on a highly complex civilization with access to the best medical facilities.

By this time, there will be few "liberals" left. Those capable of being inculcated with nihilistic, anti-natalist woke worldviews will have resigned from the gene pool, leaving only those who are, for genetic reasons, strongly traditionally religious and strongly group-oriented: the remnant pre-Industrial population.

24 Dutton and Rayner-Hilles, *The Past is a Future Country, op. cit.*

25 D. Wentland, *Identity Crisis: Teaching Imaginary Economics versus Real Economics* (London: Rowman and Littlefield, 2019), 74.

But these will be coupled with, and be very different from, the vast majority. In essence, we will come apart into fast Life History Strategists—as we breed for low IQ, genetic illness, aspects of Narcissism, Extraversion and low impulse control—and Slow Life History Strategists, as we breed for traditional religiosity, Agreeableness,[26] and conservatism. We would expect these groups to split up, based on the evidence that we get on better with those who are genetically similar to ourselves, due to this being a means of indirectly passing on more of our genes.[27] As such, even if there was the money to fund a lavish welfare system, it seems probable that there would not be the will to do so. Also, in a context of very reliable contraception, those that desire children—so long as they were reasonably genetically healthy—would be able to have them and there is evidence of a significant heritability to baby fever and to desiring lots of children.[28] Indeed, fertility itself is significantly heritable.[29] This could create a huge population spike and a consequent famine and population collapse.

Slow Life History Strategy

So, this is likely what the future holds, but what about the present? What is the current state of eugenics? As we have just seen, there is some evidence of a eugenic effect on personality. In an ecology of very low child mortality, unless you have genetic defects or cannot attract a mate, then if you want children you can have them. People who are high in Agreeableness desire children as they desire to nurture. As such, Agreeableness is weakly cor-

26 M. Jokela, "Birth-cohort effects in the association between personality and fertility," *Psychological Science,* 23 (2012): 835–841.

27 J. P. Rushton, "Ethnic Nationalism: Evolutionary Psychology and Genetic Similarity Theory," *Nations and Nationalism,* 11 (2005): 489–507.

28 A. Rotkirch, "All that she wants is a (nother) baby? Longing for children as a fertility incentive of growing importance," *Journal of Evolutionary Psychology,* 5 (2007): 89–104.

29 J. Collins and L. Page, "The heritability of fertility makes world population stabilization unlikely in the foreseeable future," *Evolution and Human Behaviour,* 40 (2019): 105–111.

related with fertility.[30] Agreeableness is a component of a broader *K*-strategy. *K*-strategists tend, all else being equal, to desire children because they want to nurture, so, overall, there is a weak correlation between aspects of *K*-strategy and fertility.[31]

In addition, both conservatism and traditional religiosity quite strongly predict fertility, and these are associated with *K*-traits, such as Agreeableness, Conscientiousness and mental stability, as well as with genetic health. On the other hand, as noted, we are selecting for fast LHS traits: low intelligence, low Conscientiousness, and aspects of psychopathology and Narcissism. And we are also selecting for Extraversion.[32]

Contraception and Abortion

Contraception is almost invariably dysgenic in terms of intelligence and in terms of Conscientiousness. People with higher intelligence and higher impulse control will be more successful users of contraception, resulting in youthful, accidental, and numerous pregnancies by females who are low in Conscientiousness and low in intelligence. On the other hand, people with low Agreeableness (who are selfish) and with high Neuroticism tend to not want to have children. Reliable contraception permits them to not have children, meaning that it helps to remove selfish people and mentally unstable people from the gene pool. So, it helps reduce unpleasantness and mental instability although it favours impulsivity and low IQ.

As many as 25 per cent of pregnancies in many Western countries are aborted, this being more than sufficient to influence the nature of the population.[33] In Britain, 20 per cent of babies

30 M. Jokela, "Birth-cohort effects in the association between personality and fertility," *op. cit.*

31 M. A. Woodley of Menie, T. Cabeza De Baca, H. Fernandes et al., "Slow and Steady Wins the Race: *K* Positively Predicts Fertility in the USA and Sweden," *Evolutionary Psychological Science*, 3 (2017): 109–117.

32 M. Jokela, "Birth-cohort effects in the association between personality and fertility," *op. cit.*

33 S. Mundasad, "Abortion study: 25% of pregnancies terminated, estimates suggest," *BBC News* (12th May 2016), https://www.bbc.com/news/health-36266873.

conceived (who are not miscarried) are aborted, based on comparing the number of live births and the number of abortions in a given year.[34] Moreover, where abortion is illegal, low IQ women become pregnant anyway, with poverty predicting unintended pregnancy in such countries,[35] so banning abortion does not act as an incentive against strongly genetic traits. Studies directly linking abortion to IQ are problematic because female subjects will tend to downplay the number of abortions that they have had, with many females having multiple abortions.[36] So, to understand the relationship with intelligence, we have to look at proxies for intelligence such as socioeconomic status.

Now, we would expect low IQ women to be more likely to get pregnant by accident as they would be more impulsive and would be less efficient users of contraception. A sound proxy for intelligence is educational attainment, which correlates with it at between 0.5 and 0.7.[37] Data from Finland—which is of excellent quality because it has a small population and keeps detailed registers—is very clear, across cohorts, on what kind of women have abortions: "Women with basic education had a higher likelihood of abortion than others and the association grew stronger for later cohorts."[38] In other words, it is the least intelligent females—with the worst academic qualifications—who are the most likely to undergo abortion. And they do so because they get pregnant by accident. Ban abortion and you compel low IQ females to have more children—as we know abortion's illegality does not cause

34 S. Bardon, "Fact check: Does one in every five pregnancies in England end in abortion?" *Irish Times* (1st May 2018), https://www.irishtimes.com/news/politics/fact-check-does-one-in-every-five-pregnancies-in-england-end-in-abortion-1.3480584.

35 J. Bearak, A. Popinchalk, B. Ganatra et al., "Unintended pregnancy and abortion by income, region, and the legal status of abortion: estimates from a comprehensive model for 1990–2019," *The Lancet*, 8 (2020): 9.

36 M. A. Woodley of Menie, J. Sanger, and G. Meisenberg, "No relationship between abortion numbers and maternal cognitive ability," *Personality and Individual Differences*, 104 (2017): 489–492.

37 Dutton and Woodley of Menie, *At Our Wits' End*, *op. cit.*

38 H. Väisänen, "The association between education and induced abortion for three cohorts of adults in Finland," *Population Studies*, 69 (2015): 373–388.

congenitally impulsive women to avoid accidental pregnancy—in a context in which there is already a weak negative correlation between fertility and intelligence.[39]

In other words, ban abortion and you accelerate the decline into chaos. Interestingly, among young women who become pregnant by accident, the ones who have an abortion have better socioeconomic life outcomes. This may be due to having fewer children, and because they are intelligent enough to think about the future and the consequences of their having children in a way that those with even lower intelligence are not.[40] But even so, it is fairly clear that very low intelligence predicts abortion overall. It is also clear from US data that low socioeconomic status strongly predicts abortion.[41] Indeed, it has been calculated that abortion in the US has resulted in a significant reduction in crime rates, by as much as 47 per cent, due to people with low IQ and psychopathic traits being more likely to abort their unwanted offspring.[42] A study of the psychiatric histories of women who had undergone abortions (which looked at their psychology long prior to their having had an abortion) found them to be "three times more likely to report a history of any mental disorder. The highest odds were found for conduct disorder and drug dependence."[43] "Conduct disorder" is, effectively, a synonym for psychopathic personality traits, with reference to those under the age of 18. "Drug dependence" implies low intelligence and low Conscientiousness, an aspect of psychopathic personality. This is consistent with evidence that Agreeableness is positively

39 Dutton and Woodley, *At Our Wits' End, op cit.*

40 D. Ferguson, J. Boden, and J. Horwood, "Abortion among young women and subsequent life outcomes," *Perspectives on Sexual and Reproductive Health,* 39 (2007): 6–12.

41 C. Dehlendorf, L. Harris, and T. Weitz, "Disparities in Abortion Rates: A Public Health Approach," *American Journal of Public Health,* 103 (2013): 1772–1779.

42 J. J. Donohue and S. D. Levitt, "The Impact of Legalized Abortion on Crime over the Last Two Decades," *American Law and Economics Review* (2020).

43 J. van Ditzhuijzen· M. Have, R. Graaf et al., "Psychiatric history of women who have had an abortion," *Journal of Psychiatric Research,* 47: (2013): 1737–1743.

associated with fertility. Altruistic people desire to nurture and to care for others, are high in empathy (so would never want to abort their baby), and thus tend to have children and, indeed, have many of them.[44]

The other predictor of pursuing abortion is mental instability. A study found that women who have undergone abortion are elevated in mental ill health—such as depression and anxiety—by around 30 per cent compared to those who have never had an abortion. Moreover, the study found that the abortion was not causal to the mental ill health. This is congruous with the study, discussed above, which found early life personality disorders among women who subsequently undergo abortion.[45] This may be because depression and anxiety predict irrational or risky behaviour, including sexual risk-taking. In addition, a pregnant woman low in these traits might feel that "everything will be okay," whereas a woman high in them would be deeply anxious about the future and strongly absorbed in her own negative feelings. All of the negative psychological traits discussed are strongly heritable, meaning that their children are likely to have them. Such children, 25 per cent of pregnancies, can be expected to have a strongly negative impact on society. In the US: "The cumulative impact of legalized abortion on crime is roughly 45%, accounting for a very substantial portion of the roughly 50–55% overall decline from the peak of crime in the early 1990s."[46] Put simply, for those who value civilization, legal abortion can be regarded as eugenic.

Clearly, both abortion and contraception impact genetic mental and physical health in other ways. Reliable contraception means that accidental offspring are less likely to be born to geriatric parents, and especially to geriatric fathers, something

44 M. Jokela, "Birth-cohort effects in the association between personality and fertility," *op. cit.*

45 D. Ferguson, J. Horwood and J. Boden, "Abortion and mental health disorders: Evidence from a 30-year longitudinal study," *British Journal of Psychiatry,* 193 (2018): 6.

46 J. Donahue and S. Levitt, "The Impact of Legalized Abortion on Crime Over the Last Two Decades," *Becker Friedman Institute* (Working Paper, 2019), https://bfi.uchicago.edu/wp-content/uploads/BFI_WP_201975.pdf.

which would reduce the percentage in populations born with high mutational load and thus with serious conditions that may stem from this. Legal abortion and pre-natal screening mean that if offspring have any serious conditions then they can be aborted and this is frequently what occurs. As already explored, the abortion rate for foetuses with Down's Syndrome has gradually increased and in Britain, by 2010, 90 per cent of foetuses with the condition were being aborted.[47] In Iceland, the abortion rate for Down's Syndrome is close to 100 per cent, though the occasional Down's Syndrome child is still born due to the screening procedure not being 100 per cent reliable.[48] Under Icelandic Law, for example, although it is not compulsory to have pre-natal screening, mothers must be told about it and told what it can achieve. In this sense, it can be argued that the state nudges them in the direction of aborting unhealthy foetuses and, thus, in the direction of eugenics. As Icelandic geneticist Kari Stefansson has put it: "We try to do as neutral counselling as possible, but some people would say that just offering the test is pointing you towards a certain direction."[49] This seems to be a fair argument. Why even offer pre-natal screening unless there is an implicit belief that most people would want to abort a foetus with serious genetic problems, in other words, would want to engage in eugenics, for the good of themselves, their child, society, or all three?

The impact of contraception is rather more ambiguous. It promotes low intelligence and poor impulse control. However, the negative correlation between education and fertility is strongest in the less developed countries, those with much poverty and income inequality and with low average IQ. The effect is less, and is observed almost only for women, in the richer countries. One interpretation would be that dysgenics invariably follows

47 *TheJournal.ie, FactCheck:* "Are 90% of babies with Down syndrome in Britain aborted?" (4th February 2018), https://www.thejournal.ie/fact-check-babies-abortion-3823611-Feb2018/.

48 J. Quinones and A. Lajka, "What kind of society do you want to live in? Inside the country where Down syndrome is disappearing," *CBSN* (14th August 2017), https://www.cbsnews.com/news/down-syndrome-iceland/.

49 Quinones and Lajka, "What kind of society do you want to live in?" *op. cit.*

fertility decline because the most rational and intelligent types, and also those who are most environmentally sensitive, are those who adopt contraception first. Long after the fertility transition, when all social classes are familiar with contraception and are increasingly undergoing the dysphoria of an evolutionary mismatch, this dysgenic effect is much attenuated. Now it is rivalled by a eugenic effect when those who live in stable marriages and who can afford to raise many children—*K*-strategists—decide to have children, while many of those living in more disorderly circumstances or in poverty, who will tend to be selfish r-strategists who are low in a desire to nurture, use contraception to prevent the children they cannot afford or do not desire.

Voluntary Euthanasia and Imprisonment

The impact of voluntary euthanasia is relatively minor, and also ambiguous. As of 2022, it is legal in Belgium, Canada, Colombia, the Netherlands, New Zealand, Spain, Switzerland, and in some Australian and US states.[50] The suffering that patients undergo has to be accepted to be unbearable and with no prospect of improvement. As of 2016, there were roughly 1,000 assisted suicides annually in Switzerland, which has a population of about 8.5 million.[51] This makes suicide easier and, perhaps, more socially acceptable. Legality may also act as a pressure on the seriously ill and helpless. On the one hand, it could be argued that the promotion of voluntary euthanasia creates a dystopian evolutionary mismatch. What kind of society—other than a moribund one—encourages its own members to kill themselves?

On the other hand, the genetically fit, with genetically adaptive inclinations, are likely to be the most resistant to the destructive calls of the sirens of dystopia. At the same time, those who

50 C. Morrison, "They're glorifying suicide: Canadian fashion company is slammed for 'dystopian' media campaign that promotes EUTHANASIA — as country prepares to open up assisted death to the mentally ill," *Mail Online* (30th November 2022).

51 N. Davis, "Euthanasia and Assisted Dying Rates Are Soaring. But Where Are They Legal? *The Guardian* (15th July 2019), https://www.theguardian.com/news/2019/jul/15/euthanasia-and-assisted-dying-rates-are-soaring-but-where-are-they-legal.

become seriously ill and who also spread negative social epistasis are likely to be more drawn to it, causing the relatively healthy to behave more adaptively in their absence. Adolescent intelligence has been found to be negatively correlated with suicidal behaviour later in life. However, in that impulsivity is an element both of low intelligence and of many suicides, it seems unlikely that voluntary euthanasia would have any impact on elevating intelligence.[52]

Some nations, especially the US, have very high rates of incarceration and keep criminals in prison for long periods of time, sometimes the rest of their lives, where there are rules such as "three strikes and you're out." Under this system, if you're convicted of a serious offence for a third time, then you will receive a life sentence.[53] In addition, many states have the death penalty for murder. Currently almost 30 per cent of African-American males spend some time in prison compared to a population average of 5 per cent, so widespread is the use of imprisonment. We have already seen that most crime, and especially most violent crime, is committed by young males with high psychopathic traits and low IQ. Consequently, mass incarceration of serious criminals can be regarded as eugenic. It prevents them breeding during the period of their lives in which they are most likely to do so, as even male fertility (and sex drive and physical attractiveness) declines with age.[54] In Finland, 5 per cent of the youth population is responsible for between 55 and 100 per cent of youth crime. The consequences of taking these people out of society before they can procreate are obvious.[55]

52 N. Bittar, D. Falkstedt, and A. Wallin, "How intelligence and emotional control are related to suicidal behavior across the life course: A register-based study with 38-year follow-up," *Physiological Medicine,* 50 (2020): 2265–2271.

53 F. Zimring, G. Hawkins, and S. Kamin, *Punishment and Democracy: Three Strikes and You're Out in California* (Oxford: Oxford University Press, 2001).

54 J. C. Oleson, "The New Eugenics: Black Hyper-Incarceration and Human Abatement," *Social Sciences,* 5 (2016): 66. "

55 J. Pitkänen, M. Aaltonen, K. Suompää et al., "Nuorisorikollisuuden esiintyvyyden ja kasautumisen muutokset syntymäkohorteissa 1986–2000," *Yhteiskuntapolitiikka,* 87 (2022): 475–490.

In Vitro Fertilization

In Vitro Fertilization is also ambiguous in terms of eugenics. It may have a very small eugenic impact, considering how few children are born employing it, by allowing more intelligent and pro-social people to have more children. First, it is used only by parents who desperately want a child. Love of children can be considered a desirable trait that we ought to select for. Second, IVF is distressing, especially for the woman, involving the use of many drugs with unpleasant side effects. Therefore, it is used only by wealthy people and by people who can cope with adversity. These are again traits that correlate with "good genes." Thirdly, we would expect, for example, women who were more intelligent and more pro-social to be more educated and career-oriented. They might, therefore, delay their fertility until it is difficult for them to become pregnant naturally. IVF would potentially permit them to pass on their genes. Fourthly, IVF is expensive, meaning that only relatively wealthy people can generally afford it and these people will tend to be relatively high in intelligence and moral character. That said, by the age of 9 there is no IQ difference between IVF children and controls.[56]

However, there are serious disadvantages to IVF. A healthy woman should be able to get pregnant, eventually, even well into her forties, and a healthy man would have no problem fertilising her. Genetically healthy people, therefore, should not require IVF or any related technology. IVF, thus, helps to perpetuate mutational load, poor genetic health, and infertility. Indeed, it might be argued that it is inherently dysgenic because, with current technology, medics can merely select a healthy-looking sperm with which to inseminate the egg. However, it is most improbable that this sperm would be the one in half-a-billion sperm that was so healthy that it out-competed all the other sperm and made it through the woman's fierce immune system to successfully impregnate her. In other words, such a sperm would almost certainly not have been able to successfully fer-

56 H. Zandstra, L. Smits, S. van Kuijk et al., "No effect of IVF culture medium on cognitive development of 9-year-old children," *Human Reproduction Open*, 4 (2018): hoy018.

tilise a healthy woman under natural conditions. In this way, IVF simply perpetuates poor genetic health and especially congenital infertility.[57] This, with our current level of technology, is also a problem with embryo selection and egg donation (insomuch as it must involve IVF of some kind).

Wokeness

A further eugenic issue may be the ideology of wokeness itself. On the one hand, it is dysgenic in the sense that when it is the dominant ideology then intelligence will assist one in convincing oneself of its veracity and, accordingly, limiting one's fertility and not pushing one's children along an adaptive road map of life which prepares them for adulthood and competition for mates, and which instils them with the need to breed. On the other hand, wokeness will push conformist people along a maladaptive, anti-natalist road map by insisting that white people like them are evil and should not breed, males are evil, the interests of other groups should be puts above one's own, one should renounce having children to save the planet, that life is pointless, and that there is no such thing as truth. Only those who can resist this extreme evolutionary mismatch—in which the leaders of your own group are directing you to destroy your group and yourself—will pass on their genes. Accordingly, those who, for genetic reasons, are conservative, group-oriented, and religious—in other words, the remnant pre-industrial population, high in genetic health—will be left, albeit alongside people whose IQ is too low to be inculcated with the ideology. In this sense, wokeness is the new "Crucible of Evolution."

Bertrand Russell, interestingly, predicted this, though with a specific focus on committed Catholics due to their high fertility in Britain and France at the time, when he discussed the issue.

> It seems unquestionable that if our economic system and our moral standards remain unchanged, there will be, in the next two or three generations, a rapid change for the worse in the character of the population

57 H. Hanevik, D. Hessen, A. Sunde, and J. Breivik, "Can IVF influence human evolution?" *Human Reproduction*, 31 (2016): 1397–1402.

> in all civilized countries, and an actual diminution in the numbers of the most civilized.
>
> The diminution in numbers will, in all likelihood, rectify itself in time through the elimination of most of the characteristics which at present lead to a small birth-rate. Men and women who still believe in the Catholic faith will have a biological advantage; gradually a race will grow up which will be impervious to all assaults of reason, and will believe imperturbably that limitation of families leads to hell-fire.[58]

Eventually those kinds of people—who will be genetically highly conservative but also not very bright—will reach a critical mass: a tipping point.

The Future of Eugenics

Richard Lynn has argued that gene therapy, germ line therapy and even cloning could be future possible ways of furthering eugenics. Gene therapy involves modifying a person's genes to treat or cure a disease. Germ line therapy—also known as "Genetic Engineering"[59]—is when DNA is transferred into reproductive cells in an attempt to cure a genetic condition. Cloning—if the technology were ever good enough to carry it out on humans—would simply involve cloning a genetically healthy individual. The problem with the first two possibilities is that they are merely a short-term solution. People that have any particular genetic disorder are likely to have general high mutational load and these technologies do nothing to deal with that. Cloning of humans, based on our current level of technology, is not viable.[60]

But what about cutting edge technologies? One possibility is germ-line gene therapy, for example in cases where both prospective parents have sickle cell disease, which is autosomal

58 Quoted in J. Day, "Bertrand Russell on Eugenics," *Mankind Quarterly,* 55 (2015): 254–267; B. Russell, *Principles of Social Reconstruction* (London: George Allen & Unwin, 1916), 180–181.

59 J. Glad, *The Future of Human Evolution: Genetics in the Twenty-First Century* (Schuylkill Haven, PA: Hermitage Publishers, 2008), 13.

60 Lynn, *Eugenics, op. cit.*

recessive. Without germ-line gene engineering, there is no way these parents will ever have a healthy child. This can be attempted although it has so far not been tried. The method for this is the CRISPR-Cas system, which has been developed for gene manipulation from a microbial immune system. It can be used to induce specific genetic changes; for example it can repair the defective gene that causes sickle cell disease, or it can induce a mutation that protects from Alzheimer's. Use of this technology for germ-line genetic engineering is considered premature at this point because there is a likely risk of inducing haphazard mutations. Still, the method may become safe enough with further improvements.[61]

Secondly, we have embryo selection. This is actually feasible with present methods. Everyone who plans on becoming a parent gets their genome sequenced. This will alert prospective parents and their doctors to risky genes that they had better not transmit to their children. Reproduction is then done via IVF. Every embryo gets its genome sequenced, polygenic scores are calculated and so on. Then the best embryos are used and the rest are discarded. IVF costs about $15,000, a genome sequence costs roughly $1000, so the whole cost for an IVF cycle with embryo selection would add up to $30,000. Poor couples could not afford this. Therefore, this could be practiced on a larger scale only in a health care system like in Britain where healthcare is guaranteed for everyone, not in the American kind of system where those who cannot afford health insurance are excluded.[62]

The third strategy for effective eugenics would be a professional mother program, where embryos are generated from the sperm and egg of highly selected donors, tested for absence of genetic flaws, and then implanted in women who are paid to bear and raise the offspring. This could provide for the livelihoods of poor women who have few options on the labour market, and who like children, or older couples who like children but cannot

61 See C. Brokowski, M. Pollack and R. Pollack, "Cutting Eugenics Out of CRISPR-Cas9," *Ethics in Biology, Engineering and Medicine*, 6 (2015): 263–279.

62 S. Wilkinson, "Eugenics, embryo selection, and the Equal Value Principle," *Clinical Ethics*, 1 (2006): 46–51.

have them anymore.

Many ethicists will of course be horrified by all of these proposals, but from a consequentialist point of view it might be argued that everyone gains. Some ethicists even argue that based on the "equal value principle" it is *a priori* wrong to discriminate between healthy and disabled foetuses.[63] One wonders if such ethicists apply this principle to their lives more generally and, thus, fail to discriminate between that which is beneficial to them and that which is damaging to them. Of course, they would not and, thus, this argument fails the "test of pragmatism;" that one must be able to live by what one asserts to be correct.[64] Moreover, by the same logic, we should not attempt to cure the sick, or inoculate children against polio, as to do so privileges "health" over "sickness" or "mobility" over paralysis."[65] I would suggest that, in general, philosophical "ethics" suffers from serious problems. Following the thinking of "Non-Cognitivists," I would aver that it, essentially, takes imperatives and expressions of emotion and attempts to dress these up as "truths." But such claims are incapable of being "true." Ethical truths can only be agreed on from a shared "vantage point" and if your interlocutor will not accept the importance of maintaining civilization, for example, then there will be no arguing with him, although you might highlight his inconsistency in living off the fruits of civilization. No "vantage point" can inherently be "privileged" over another, so we simply end up with open-ended arguments in which "ethicists" are permitted to make a living by attempting to appear profound.[66]

The Problems with Eugenics

I do not see much of a future for eugenics. The problem is that

63 Wilkinson, "Eugenics, embryo selection, and the Equal Value Principle," *op. cit.*

64 W. James, *Pragmatism: A New Name for Some Old Ways of Thinking* (New York: Longmans, Green and Co., 1907).

65 J. Glad, *The Future of Human Evolution, op. cit.*, 29.

66 E. Westacott, "Moral Relativism," *Internet Encyclopedia of Philosophy* (2022), https://iep.utm.edu/moral-re/.

most parents are too careless to go much out of their way to improve their chances of a healthy child, and that they are too superstitious, either based on religious-conservative notions that we "shouldn't play God" or based on the woke rhetoric that the belief in genes is inherently evil because it implicitly opposes "equality." The worst problem is those with high IQ as this actually predicts not wanting to have children in Western samples.[67]

Let us begin with the problems with "Classical Eugenics"—promoting breeding among the more intelligent and discouraging it, or even preventing it, among the less intelligent. IQ is currently falling by as much as 2.4 points every decade in Western countries, based on a meta-analysis,[68] though some studies find that it is only around a point per decade or even less than this. Analyses of the relationship between fertility and IQ, taking into account the cumulative way in which intelligence declines, have estimated that the IQ of native Westerners, if set at 100 now, will be 85 by the end of the century.[69] So unless some major innovation takes place that somehow boosts IQ, a lot of people would have to be prevented from breeding simply to halt the decline. Assuming 30 years to a generation, anybody with an IQ of below 92 (so nearly a third of the population) would have to be prevented from breeding.[70]

In practice, this would likely set off some kind of revolution but let's focus on theory. In theory, this could work, and in that intelligence is weakly genetically correlated with physical and mental health, as already discussed, such a system would promote a general "fitness factor." A potential problem is mutational load, which negatively impacts intelligence via interfering with brain functioning,[71] building-up more quickly than eugenic

67 S. Kanazawa, *The Intelligence Paradox: Why the Intelligent Choice Isn't Always the Smart One* (Hoboken, NJ: John Wiley and Sons, 2012).

68 E. Dutton, D. van der Linden and R. Lynn, "The Negative Flynn Effect: A Systematic Review," *Intelligence,* 59 (2016): 163–169.

69 Dutton and Rayner-Hilles, *The Past is a Future Country, op. cit.*

70 I am grateful to Mr Lewis Low for providing me with these calculations.

71 M. A. Woodley and H. Fernandes, "The secular decline in general intelligence from decreasing developmental stability: Theoretical and em-

measures on intelligence are increasing fitness.

Re-imposing High Child Mortality

There are two ways this could be solved. Rising mutational load could only be truly dealt with if a child mortality rate of about 50 per cent was reintroduced. But even this might not be good enough to maintain genetic health. In eighteenth century rural Sweden, life-expectancy—once you survived childhood—was 72 years and only around 15 per cent of the adult population failed to breed, either due to infertility (2–4 per cent) or simply failure to find a mate.[72] However, other studies, such as in seventeenth century England, have found that 20 per cent of marriages were childless, partly due to high child mortality.[73] By the end of the seventeenth century, roughly 20 per cent of males and females over the age of 40 were unmarried, this being a relatively high point of not marrying in pre-Industrial England.[74] The above estimates imply that, in England, by the end of the seventeenth century, as many as 90 per cent of people who were born did not pass on their genes. It has been calculated, using computer modelling, that for a population to maintain optimum genetic health then, based on the frequency of genetic mutations, approximately 88 per cent of that population, in any given generation, should fail to breed.[75] It seems unlikely that any government that imposed such a system could survive.

An alternative would be to breed for a general fitness factor that included intelligence. In animal husbandry, this is known

pirical considerations," *Personality and Individual Differences,* 92 (2016): 194–199.

72 T. Volk and J. Atkinson, "Is Child Death the Crucible of Human Evolution?" *Journal of Social, Evolutionary, and Cultural Psychology,* 2 (2008): 103–116.

73 R. Grassby, *The Business Community of Seventeenth-Century England* (Cambridge: Cambridge University Press, 2002), 132.

74 E. A. Wrigley and R. S. Schofield, *The Population History of England, 1541–1871* (Cambridge: Cambridge University Press, 1981), 264.

75 Y. Lesecque, P. Keightley, and A. Eyre-Walker, "A Resolution of the Mutation Load Paradox in Humans," *Genetics,* 191 (2012): 1321–1330.

as a "Genetic Breeding Value."[76] Due to these superstitions and because of people's disinterest in the welfare of future generations, in an increasingly individualistic society, we would need a dictator to impose eugenic measures, of the kind outlined above. This was needed to a lesser extent in the 1920s, because people were more group-oriented and more trusting in their leaders and thinkers, but even then there was dissent. We would, additionally, need a society that was highly group-oriented but, crucially, *not* traditionally religious, as all major religions, even Hinduism, oppose eugenic measures, such as abortion.[77] This is a serious problem, because religiousness seems to be strongly associated with believing in the inherent sanctity of life.

Furthermore, as intelligence, trust and cooperation decline, this kind of society would be increasingly difficult to achieve. Such a dictator would have to force the majority of the population to not have children, or compel them to submit to a state-mandated system in order to have children, such as the rigged lottery proposed by Plato. It seems obvious that the result would be degeneration into conflict and the collapse of the entire system. One can imagine that those who lost the lottery would, correctly, start to believe in a "conspiracy theory" which was preventing them from winning. Even in 1931, the Eugenics Society recognised that "the religious and moral objections" to sterilisation "are so strong in the minds of many people as to render the advocacy of compulsion in the matter inadvisable."[78] Indeed, even if positive eugenics was employed, one can imagine those who did not receive the relevant tax breaks developing a conspiracy whereby they were in fact super-humans being oppressed by the frightened sub-human elite.

Moreover, even putting this aside, the very people with high

76 M. Cortellari, A. Negro, A. Bionda et al., "Using Pedigree and Genomic Data toward Better Management of Inbreeding in Italian Dairy Sheep and Goat Breeds," *Animals*, 12 (2022): 2828.

77 K. Aramesh, "Perspectives of Hinduism and Zoroastrianism on abortion: a comparative study between two pro-life ancient sisters," *Journal of Medical Ethics and History of Medicine*, 12 (2019): 9.

78 *Yorkshire Post and Leeds Intelligencer*, "Eugenic Sterilisation" (29th December 1931).

IQ who are breeding—and thus likely to eventually rise to positions of power—are the traditionally religious and they tend to strongly oppose abortion and other forms of eugenics, as noted above. To make matters worse, people don't even seem to desire highly intelligent children. They want children that are well-adjusted and content, which highly intelligent children are often not, not least due to sometimes being high in certain autism traits,[79] an issue upon which we will expand below. This can be seen in the preferences of the recipients of donor sperm, among whom a variety of traits are desired, including intelligence but also similarity to themselves. They would have to be compelled to focus on what the government regarded as desirable.[80] So, it seems most unlikely that Classical Eugenics could be imposed.

Would It Even Work?

Moreover, even if it could be pursued, it probably would not accomplish what it desires to, in the sense of a avoiding an eventual collapse. As people become more intelligent, the positive manifold between the different components of intelligence becomes weaker, meaning that highly intelligent people can be very bad at things that are weakly associated with intelligence; in other words, highly impractical.[81] This has been found to be true of many scientific geniuses—innovators of widely recognised major breakthroughs—who, despite their brilliance, cannot cook or drive a car.[82] As people become more intelligent, they become higher in Openness-Intellect—openness to new things and

79 T. Nishyama, H. Taniai, T. Miyachi et al., "Genetic correlation between autistic traits and IQ in a population-based sample of twins with autism spectrum disorders (ASDs)," *Journal of Human Genetics*, 54 (2009): 56–61.

80 C. Drewes, *Anonymous Sperm Donor Preferences of Non-Genetic Mothers* (Master's Thesis. Smith College, MA, 2009), 78.

81 D. Blum and H. Holling, "Spearman's law of diminishing returns: A meta-analysis," *Intelligence*, 65 (2017): 60–66.

82 E. Dutton and B. G. Charlton, *The Genius Famine* (Buckingham: University of Buckingham Press, 2015).

interest in ideas, this correlates with IQ at 0.3[83]—and lower in instinct, including, it seems, the instinct to have children, especially if they live in an evolutionary mismatch, being highly environmentally sensitive.

Very high intelligence is associated with autistic traits, including low empathy, poor social skills and becoming easily overwhelmed by stimuli.[84] Alleles for autism overlap with alleles for very high intelligence.[85] This seems to be because autistic traits—systematizing and taking in more information—allow you to solve problems better. However, this implies that there is, in effect, a limit to human intelligence. Beyond a certain level, intelligence becomes fitness-reducing, because, in mismatched circumstances, it stops you from breeding, causing civilization to collapse or at least shrink substantially but humanity to continue. And this is good, because if this didn't happen, and we went on becoming more intelligent forever, then society would be full of impractical autistics with no social skills and no desire to breed, and humanity would die out. Nevertheless, average intelligence would have to be much higher than it is now—and much higher than it was in 1800, when average IQ may have been around 120 or more—for this to be a serious problem. We could get round these problems via some form of genetic engineering.

In theory at least, average IQ could be raised significantly in parallel with health, though it may lead, ultimately, to a fertility collapse and, consequently, a general collapse. However, if some kind of eugenics policy was imposed then we could at least slow down our decline and, perhaps, preserve more knowledge to be rediscovered by the next civilization, much as our civilization rediscovered Classical texts, preserved in Byzantium, during the

83 D. Nettle, *Personality: What Makes You Who You Are* (Oxford: Oxford University Press, 2007).

84 T. Nishyama, H. Taniai, T. Miyachi et al., "Genetic correlation between autistic traits and IQ in a population-based sample of twins with autism spectrum disorders (ASDs)," *Journal of Human Genetics,* 54 (2009): 56–61.

85 B. Crespi, "Autism as a Disorder of High Intelligence," *Frontiers in Neuropsychology,* 10 (2016): 300.

Renaissance.[86]

However, this raises another problem. As we have seen, geniuses tend to combine outlier high IQ with traits that are associated with poor fitness, such as autism and aspects of psychopathology. An optimum low level of them stays in the population, possibly due to group-selection. Without this, we would be breeding a people who, though highly intelligent, would not tend to make great innovations, as the genetic diversity which permitted the unlikely gene combinations that permit this would be bred out. In fact, as we discussed earlier, a functioning complex society—where people must pursue a variety of occupations optimally well—would require diversity, at least in terms of personality; in effect, a series of psychologically different castes. So, once more, we would need to turn to genetic engineering.

Alternatives to Classical Eugenics

We have seen that Classical Eugenics suffers from serious problems, which could, in theory, be solved via genetic engineering. How viable is this possibility? We are currently at a very elementary stage of gene therapy (in a society of rapidly declining IQ) where we can disable only single mutant genes.[87] Intelligence is massively polygenic—it involves thousands of genes working together—making it very difficult to know how any individual gene that gene-editing might target will impact the whole or what negative side effects there might be from interfering with any particular gene. William Hamilton, Royal Society Research Professor at Oxford University, whom we met earlier, noted that we might think that getting rid of the allele that causes sickle cell anemia is a brilliant idea, but we'd also be eliminating resistance to malaria. "Might this prove to be only the tip of an iceberg of

86 J. Leonhart, *Latin: Story of a World Language* (Cambridge, MA: Harvard University Press, 2013), 200.

87 C. Funk, A. Tyson, B. Kennedy and C. Johnson, "Biotechnology Research Viewed With Caution Globally, but Most Support Gene Editing for Babies To Treat Disease," *Pew Research Center* (10th December 2020), https://www.pewresearch.org/science/2020/12/10/biotechnology-research-viewed-with-caution-globally-but-most-support-gene-editing-for-babies-to-treat-disease/.

similar examples?" he asks.[88] As American philosopher F. Roger Devlin has put it:

> A person ignorant of viticulture would probably assume that the best wine would be produced by the best soils. In fact, when vines are planted in especially rich soil, the result is an enormous profusion of leaves and very few grapes. High-quality wine grapes are grown in relatively weak—but not *too* weak—soil: it is precisely the vine's struggle against a suboptimal environment which brings out the best in grapes [...] This, incidentally, is what makes eugenic intervention such a delicate matter: nature may be acting more wisely than any possible human agent by allowing certain normally sub-optimal traits to survive within a population.[89]

Following this reasoning, it is probable that any eugenic intervention will simply make things worse. In this regard, American philosopher David Skrbina has argued:

> As we are painfully learning, the 4-billion-year evolution of the Earth has resulted in a very efficient, very finely-tuned planetary system. The Earth (prior to the advent of modern man) makes optimal use of every available resource. Every waste product is used or hidden away. Every trade-off is balanced and accounted for. The global bio-system, as a whole, is at its peak "productivity." The total production of animal life, food, fruit, seeds and grains, is the highest that can be sustained over the long run. Any action by mankind outside of his aboriginal condition must therefore result in a net diminishment of the global system.[90]

The evolution of humans has occurred in a similarly fine-tuned

88 W. D. Hamilton, *Narrow Roads of Gene Land: Last Words* (Oxford: Oxford University Press, 2005), 213–214.

89 F. R. Devlin, "Premature Birth and Genius," *Unz Review* (9th February 2022), https://www.unz.com/article/premature-birth-and-genius/.

90 D. Skrbina, "Technological Anarchism: Reconsidering the Unabomber," *Chromatikon*, 5 (2009): 191–201.

way, and it might be argued that intermittent population collapses are part of this. Interfering with it is almost certain to have unforeseen negative consequences, such as an eventual absolutely catastrophic collapse and even extinction of the species. Further, the pure unnaturalness of such a system would likely create an even more pronounced evolutionary mismatch and even greater dysphoria, which would itself have unforeseen negative consequences. To interfere would be akin to putting out intermittent forest fires, which are necessary for the ecosystem to flourish. The eventual result is an absolutely catastrophic forest fire that is beyond our control. With regard to interference in evolutionary affairs, as British Prime Minister the Marquess of Salisbury (1830–1903) maintained: "Whatever happens will be for the worse, and therefore it is in our interest that as little should happen as possible."[91] This all implies that the only realistic possibility is Classical Eugenics, but we have already discussed the problems with that.

Fermi's Paradox

It has been argued that civilizational collapse—as something we simply cannot stop—neatly explains Fermi's Paradox, whereby, statistically, we would expect there to be advanced alien life, yet there is no evidence of it. If civilizations always collapse before reaching sufficient intelligence and thus technological innovation to find other alien civilizations, then Fermi's Paradox is no longer a paradox.[92] That said, it might be argued that this is only half the explanation of Fermi's paradox. The other half is that once a species passes the critical threshold where it starts controlling its own evolution (which I would define as the vital difference between an intelligent species and an ordinary animal species), it will raise its own intelligence, and with it the desire for knowledge and understanding. At some point, knowledge, understanding and evolutionary mismatch will be so high that

91 R. Taylor, *Lord Salisbury* (London: Lane, 1975), 139.

92 See, M. Sarraf, M. A. Woodley of Menie and C. Feltham, *Modernity and Cultural Decline: A Biobehavioral Perspective* (London: Palgrave Macmillan, 2019).

the members of this species come to agree that life is pointless, for whatever direct or indirect reason, and it is best to stop reproducing and to die out.

Another possibility is that these intelligent beings keep improving themselves, but reach a limit that they understand they can never overcome because of their biological limitation. This would most plausibly be the inability to make oneself immortal. In consequence, these highly intelligent beings may decide to create an artificial life form in their own image, with exactly the species-typical psychology but endowed with immortality. They would then end their own existence, being replaced by this immortal replica of themselves. If we are indeed visited by UFOs, these aliens will be either a self-improved species that has long abandoned the old dependence on mutation and selection, or an artificial replica that has replaced a bygone organic species. Either way, they would immediately recognize us as common animals, not an intelligent species capable of directing our own evolution.

Would Civilization Really Collapse?

It might be argued that civilization wouldn't collapse with an IQ of 85, despite the fact that, based on per capita innovation rates, this is what the average IQ of Western countries was during the Dark Ages.[93] It might be suggested that many relatively complex countries exist today such as Brazil that have an IQ of about 85.[94] Accordingly, there is time for a eugenic regime to develop, in which to gain sufficient understanding of genetics in order to perfect it. However, even putting aside the fact that Brazil and countries of a similar IQ rely on aid from developed countries,[95] this misunderstands the globalised and hierarchical nature of the

93 Dutton and Rayner-Hilles, *The Past is a Future Country, op. cit.*

94 R. Lynn and D. Becker, *The Intelligence of Nations* (London: Ulster Institute for Social Research, 2019).

95 *Reuters,* "Brazil seeks $1 bln in foreign aid to curb Amazon deforestation by 30–40% -environmental minister" (8th April 2021), https://www.reuters.com/business/environment/brazil-seeks-1-bln-foreign-aid-curb-amazon-deforestation-by-30-40-environmental-2021-04-08/.

modern economy.

The whole world depends on the economies of scale created by large first world institutions, gradually set up over many decades by a tiny minority of engineers and managers, when compared to the whole world population. A country like Brazil greatly depends on a densely clustered cognitive elite that emerged out of places like the San Francisco Bay area. Their gradual dissolution actually threatens the developing world more than the immediate vicinity because the complex systems which they have enacted, which allow developing countries to just about remain technological, will break down.[96] In this sense, not all intelligence is located squarely inside people's heads. A lot of it is implicitly there, present in complex organisations and trade networks. This explains why the correlation between IQ and earnings is only about 0.3, but the correlation between national IQ and GDP is about 0.8.[97]

"The Elites Won't Let a Collapse Happen."

A related argument is that "the elites" will stop society collapsing, by changing the conditions that are causing dysgenics. They will gradually introduce "transhumanism," creating a semi-robotic elite, as Israeli historian Yuval Harari has predicted.[98] But these conditions have been known for centuries, and this supposedly all-powerful elite surely knows what's happening, yet they have done nothing. Such people will exploit the collapse and ensure that they are insulated from the anarchy—like African despots—but they show no evidence of doing anything about actually preventing tomorrow's disorder from unfolding. Moreover, the elites will be becoming less intelligent as society does, and have long been doing so. Thus, they are increasingly unable to do anything

96 See, H. Rindermann, *Cognitive Capitalism: Human Capital and the Wellbeing of Nations* (Cambridge: Cambridge University Press, 2018).

97 G. Francis and E. Kirkegaard, "National Intelligence and Economic Growth: A Bayesian Update," *Mankind Quarterly*, 63: 1; R. Lynn and T. Vanhanen, *Intelligence: A Unifying Construct for the Social Sciences* (London: Ulster Institute for Social Research, 2012).

98 Y. Harari, *Homo Deus: A Brief History of Tomorrow* (New York: Random House, 2016).

about it and they are decreasingly long-term in their thinking.

But, anyway, which is more likely? Are we unique, unlike any other civilization, meaning we will not collapse due to a conspiracy of highly intelligent psychopaths who are working together to create a super-race and enslave the rest of us, there somehow being some impenetrable border between "us" and this "elite"? Or, will we go the way of every other civilization, including Roman and Middle Eastern Muslim civilization? If we were going the way of Roman civilization then we would notice evidence of declining intelligence, collapsing traditional religiosity, feminism, multiculturalism, materialism, collapsing traditional moral standards, a loss of harsh martial values and their replacement with decadence in the form of self-realisation or compassion, polarisation, the debasement of democracy, despotism, and a general coming apart of formerly large polities. In that this is actually what is happening and there is solid evidence for it, it seems a much more plausible state of affairs.

While this was happening, we would expect people, due to decreasing intelligence (and thus decreasing logical abilities and also increasing dominance by cognitive biases, such as over-detecting agency) and increasing schizophrenia and general paranoia (due to higher mutational load) to be arguing that there was a Satanic elite that was going to succeed in enslaving us all. But the elite, failing to breed sufficiently, would fall, as it always has done before. In general, if one is propounding a supposedly logical argument that very strongly reflects an evolved cognitive bias—such as over-detecting agency—one should very seriously question this argument. Indeed, there is evidence that we are subject to a cognitive bias whereby if we can imagine something being different then we will imagine it being better, which may be why so many people assume that civilization simply cannot collapse this time around.[99]

As for the conspiracy element, belief in in conspiracy theories that cannot theoretically be falsified—and where other explanations are far more likely—is associated with deficiencies in ana-

99 A. Mastroianni and E. Ludwin-Peery, "Things Could Be Better" (2022), https://psyarxiv.com/2uxwk.

lytic ability,[100] Narcissistic traits (perhaps the conspiracy explains why one is not as successful as one believes one should be),[101] and, as we noted earlier, paranoia and low Agreeableness.[102] The rise of conspiratorial thinking reflects an increasingly low trust society, and in this sense it is rational. But when it comes to believing in massive, large-scale, clandestine evil plots that are behind all or most important historical events, such a belief is irrational because the more people there are involved, the more inevitable is a leak. Furthermore, it can be demonstrated that such events are components of general processes and are not caused by deliberate agency. When you, even so, take the conspiratorial line, this is irrational.[103] In this sense, the growing popularity of such theories is, in part, an expression of dysgenics, combined with low IQ and paranoid people having access to large amounts of information, via the internet, and being able to easily find each other. On the other hand, there may be an evolutionary rationale to perceiving your empowered political opponents—who may be quite genetically different from you—as evil, strongly motivating you to overthrow them and take power yourself. This is a kind of high negative ethnocentrism, which is strongly adaptive in terms of group selection.[104]

The Grip of Individualism

A further problem with instituting eugenics is that we are so in-

100 V. Swami, M. Voracek, S. Stieger et al., "Analytic thinking reduces belief in conspiracy theories," *Cognition.* 133 (2014): 572–585.

101 A. Alsuhibani, M. Shevlin, D. Freeman et al., "Why conspiracy theorists are not always paranoid: Conspiracy theories and paranoia form separate factors with distinct psychological predictors," *PLoS ONE*, 17(2022): e0259053.

102 A. Goreis and M. Voracek, "A Systematic Review and Meta-Analysis of Psychological Research on Conspiracy Beliefs: Field Characteristics, Measurement Instruments, and Associations with Personality Traits," *Frontiers in Psychology* (2019).

103 See, D. Aaronovitch, *Voodoo Histories: How Conspiracy Theory Has Shaped Modern History* (London: Vintage Books, 2010), Ch. 1.

104 R. Hammond and R. Axelrod, "The evolution of ethnocentric behaviour," *Journal of Conflict Resolution,* 50 (2006): 1–11.

dividualist in our thinking that we have reached a point of runaway individualism, due to people trying to out-compete each other in ever more extreme expressions of individualism in order to gain status and acclaim in a society where this is considered virtuous. We can now define ourselves, state what gender we are, and expect others to accept this because we say so. We cannot yet define our own "race" because the woke perceive blacks as more disempowered and therefore more deserving than Whites. Being able to choose one's race, the way people are encouraged to choose their gender, is unacceptable because it would enable Whites, who are intrinsically undeserving, to get off the hook by claiming a Black identity at will, though race becomes a "social construct" the moment woke ideology is critiqued by scientists. This inconsistency is acceptable because leftists are Machiavellian, pursuing power by fake concern with the supposedly disempowered. Truth and logic don't matter. All knowledge is socially constructed, and power alone determines whose truth prevails. However, transsexuals are more disempowered than women, so they are allowed to change sex. This, however, demonstrates that we are far too individualistic to seriously do anything for the good of the group, such as trying to slow down our intellectual and genetic decline.

Above all, this kind of ideology can only be the product of dysgenics, in the way that it combines lack of intelligent reasoning with rampant Dark Triad traits. This includes a half-demented audience that can be manipulated to believe in a power-hungry "elite" that somehow are in control of everything. Of course, this is actually a reflection of low-trust combined with an inability to overcome certain evolved cognitive biases such as agency over-detection and over-detection of causality. It is the high instinctiveness, and low trust, that is seen among those who have low intelligence. The fervour with which such views are held—whether the conspiracy is systemic racism or "the elites"—is also a reflection of low IQ; though these views can also be found among more intelligent people who are high in certain Neurotic or psychopathic traits that overwhelm their rational thinking, as already explored.

We Are All Eugenicists

Perhaps the emotion surrounding eugenics is such because, for the left, it triggers "cognitive dissonance." It confronts them with an aspect of themselves that, as people who have a need to believe they are of superior moral status, they would rather didn't exist. Some of them may have had abortions, for example. To get round this problem, leftist science writer Adam Rutherford defines eugenics as "designed to sculpt society through selective breeding," as contrasted with "medical interventions designed to offer options to parents, enabling them to make choices about the medical health of lives they may want to bring into the world."[105] Thus, those who have an abortion because the child is disabled are not eugenicists. However, surely, even being offered these "choices" cannot be divorced from promoting "desirable characteristics."

Moreover, we are all products of an often quite deliberate process of eugenics. Most readers of this book will be university graduates and probably graduates in academic subjects, just like myself. The reason we were able to graduate in these academic subjects was due to a combination of relatively high intelligence, which predicts about 50 per cent of the variance in education level,[106] and personality factors such as Conscientiousness. Intelligence is around 80 per cent genetic among adults and personality at least 50 per cent so.[107] Other reasons for our ability to be graduates will include, in many cases, our parents' relatively high socioeconomic status, which is associated with their children having high intelligence,[108] and factors such as not spending our childhoods debilitated by illness. And this happened because of "eugenic" decisions by our parents.

As already discussed, females, to a greater extent than males,

105 A. Rutherford, *Control: The Dark History and Troubling Present of Eugenics* (London: Orion, 2022).

106 A. Jensen, *The* g *Factor: The Science of Mental Ability* (Westport, CT: Praeger, 1998).

107 See R. Lynn, *Dysgenics: Genetic Deterioration in Modern Populations* (London: Ulster Institute for Social Research, 2011).

108 Dutton and Woodley of Menie, *At Our Wits' End, op. cit.*

are attracted to qualities such as intelligence and Conscientiousness in potential partners. The sensible strategy for men is to simply have multiple sexual partners and specifically select for those who are young (and thus more fertile) and who are physically attractive, because this tends to indicate that they have "good genes." They are also interested in a potential partner's psychology, of course. Most obviously, if they are going to invest in the partner and her offspring then they want to be sure the offspring are theirs, so her intelligence and personality mattered even centuries ago when she was unlikely to work. They want a woman who is trustworthy, for example. Similarly, the males' looks matter to the female. She would rather not have a genetically sick husband or one who can't protect her if there's a fight, although these qualities will be more important to some women than others, such as women living in a dangerous and unstable environment. But, in general, it is true that women go for status and men go for looks and youth. Furthermore, as discussed above, people are attracted to those who are genetically similar to themselves, as this optimises the extent to which they pass on their genes. Thus, intelligent people will be attracted to other intelligent people, and may even trade "looks" or "status" in order to attain such a genetically similar mate.[109] American geneticist Kathryn Paige Harden (b. 1983)[110] refers to there being a "Genetic Lottery."[111] But it is not a "lottery." It is predictable: people with desirable genetic qualities will tend to mate with each other, producing offspring with desirable genetic qualities.

So, unless your parents (or my parents) both made a point of marrying somebody whom they found psychologically and physically repellent and whom they found in no way sexually attractive then you are a product of eugenics. To the extent that your parents thought things like, "She's quite good looking!" or "He seems really clever!" this eugenics was deliberate and

109 J. P. Rushton, "Ethnic Nationalism: Evolutionary Psychology and Genetic Similarity Theory," *Nations and Nationalism,* 11 (2005): 489–507.

110 G. Lewis-Kraus, "Can Progressives Be Convinced That Genetics Matters?" *The New Yorker* (6th September 2021).

111 K. P. Harden, *The Genetic Lottery: Why DNA Matters for Social Equality* (Princeton, NJ: Princeton University Press, 2021).

planned. To the extent that eugenics critic Adam Rutherford's mother thought, of his father, "He's a college lecturer in social psychology![112] He has reasonably high social status and must be pretty clever!" and the father thought, "She's pretty!" this is deliberate eugenics. As discussed earlier, eugenics critics Adam Rutherford and Angela Saini are both products of numerous generations of religiously-sanctioned eugenics by virtue of being part high-caste Hindu and almost certainly wouldn't be writing their popular science books if they were not.

Perhaps, in addition to growing individualism, part of the revulsion for eugenics is that people, increasingly, do not have any connection to producing food through farming, or even gardening and growing their own vegetables the way their ancestors did until recently. If they did, they could be successful only by making harsh "eugenic" decisions all the time: do away with the runt; collect your seeds from the healthiest and most productive plants. "Eugenics," as the word is commonly used, simply transfers these techniques over to humans.

Advanced Apes

The term "Advanced Ape" is very telling. It raises a fundamental question to those who believe that "this time round" things are going to be different; that "this time round" there will be no civilizational collapse or population decline. The populations of all species are subject to Malthusian Cycles, as proposed with regard to humans by English vicar Thomas Malthus (1766–1834).[113] Under these cycles, human populations increase in favourable conditions, become unsustainably high (especially when environmental conditions become less favourable), and collapse due to starvation and attendant outbreaks of disease. The process then begins all over again. This same process occurs in all species. The population, across natural history, hovers around a natural equilibrium, but there are "humps and bumps": rises and collapses.

112 *Bulletin of the British Psychological Society* (1970), 83.

113 T. Malthus, *An Essay on the Principle of Population* (London: J. Johnson, 1798).

It can be argued that it is different in this civilizational cycle because the Industrial Revolution, for the first time, permitted the rate of production to outpace population growth, leading to an enormous human population. However, this has occurred because humans have created for themselves a period of extreme ease and, thus, very weak Darwinian selection pressures. The result of that is a population that is increasingly genetically sick, genetically of low intelligence, genetically of anti-social personality, and one in which most people have lost any practical knowledge of how to attain food through hunting, gardening, farming, or foraging, and in which some people lack the desire for, or actively oppose, the survival of their group or even of humanity.

A larger percentage of the human population than ever before is completely reliant on the innovations of civilization in order to survive. Eventually, as happens to animals with runaway populations, the environment changes. Civilization will decline due to falling IQ or it will fall very quickly due, for example, to a volcanic eruption creating a nuclear winter and a desperate scramble for food in a context of the population suddenly being too high. The result will be warfare, chaos, and mass death, as occurred during the Late Bronze Age Collapse of around 1200 BC in which European civilization barely held on and all European cities were lost. It took almost a millennium of harsh Darwinian conditions in order for Europe to return to Late Bronze Age levels of civilization, and this was in a context in which civilization was not as inter-connected as it is today and in which civilization in Egypt had just about held on.[114] Egypt managed to repel the sea peoples (the remnants of which were most likely the Philistines). As populations become denser, there will be more novel epidemics, they will have greater impact, and as intelligence declines we'll be decreasingly able to develop vaccines or cures for them or persuade low-trust populations to take the vaccines or medicines. In this sense, the Industrial Revolution has merely postponed the collapse; it has meant that it will occur to a much larger and sicker population, meaning that it will be unprecedented in its ferocity and extent. War and migration will spread these diseases

114 S. Chissanthos, *Warfare in the Ancient World: From the Bronze Age to the Fall of Rome* (Westport, CT: Praeger, 2008).

and our weakened immune system will be hit by them all at once. This would potentially cause an apocalyptic collapse.

The Great Justinian Plague

Such collapses have occurred many times before, but let us look at one of the most recent to gain some kind of purchase on what such a collapse would be like. By the sixth century, the Roman Empire was already in decline. The parallels with our own situation are numerous. For many centuries, per capita innovation had been falling, but there was a relatively advanced society in many areas: sewers, clean water, durable buildings, some understanding of medicine, animal breeding that ensured large animals that could feed larger populations, and an excess of food which fed a growing Roman underclass for whom there was no work, this work being done more cheaply by foreigners.[115] The period was also warm—the Roman Warm Period—meaning that Darwinian selection pressures were weak and a larger, less group-oriented and less fit population could expand.[116] Rome, by this time, had long been doubting its traditional religiousness that promoted patriarchy and ethnocentrism as the will of the gods. There were females, and especially immigrant non-Romans, in positions of influence. The Senate had even asked the Emperor Augustus, hundreds of years earlier, how female influence could be reversed.[117] Rome was coming apart. It was increasingly politically unstable and polarised, as has been demonstrated to be the case in the West today, with the West currently being highly polarised based on various measures.[118]

At about this point, the Roman Warm Period came to a halt. It started to become colder. Darwinian selection pressure returned

115 W. Baker, *Endless Money: The Moral Hazards of Socialism* (Hoboken, NJ: John Wiley & Sons, 2009).

116 P. Nunn, *Climate, Environment, and Society in the Pacific During the Last Millennium* (Amsterdam: Elsevier, 2007), 9.

117 J. Glubb, *The Fate of Empires and the Search for Survival* (Edinburgh: William Blackwood & Sons, 1976).

118 P. Turchin, *Ages of Discord: A Structural-Demographic Analysis of American History* (Mansfield, CT: Beresta Books, 2016).

with a vengeance. Insufficient food could now be grown to feed the population, and people began to starve. The cities were increasingly abandoned as there was no longer excess food available for a society to have so many people who weren't directly associated with growing food. Buildings were then cannibalised for useful materials, before being allowed to fall into ruin.[119] These people's skills, unless they could be turned to something of more practical use to survival, were increasingly useless, and many of them may well have starved as a result. They were the PhDs in Feminist Studies of their day. The underclass starved, and so would have begun stealing and fighting, leading to a collapse in law and order. Wealthier people, with better access to food, paid violent gangs to protect them from the growing disorder even as early as the third century.[120] Large numbers of people abandoned formerly lush and bountiful farms in order to search for new pastures, leading to conflict and warfare.[121] As it got colder, any plague would be better able to spread through the already malnourished and epidemic-ridden population, which had ballooned in size during the warm period. Then, between 541 and 549, came the Justinian Plague, named after the Byzantine Roman Emperor at the time. Believed to be the same disease as the Black Death, the result was an absolutely catastrophic population collapse of about 60 per cent throughout Europe.[122] The devastation this wrought was first recorded by Procopius (c.500–565), who witnessed what happened in Byzantium:

> Now in the beginning each man attended to the burial of the dead of his own house, and these they threw even into the tombs of others, either escaping

119 D. Russo, *Town Origins and Development in Early England, C.400–950 A.D.* (Westport, CT: Greenwood Press, 1998), 79.

120 J. White, *The Roman Emperor Aurelian: Restorer of the World* (Barnsley: Pen and Sword, 2015).

121 D. Stathakopoulos, *Famine and Pestilence in the Late Roman and Early Byzantine Empire: A Systematic Survey of Subsistence Crises and Epidemics* (London; Routledge, 2017).

122 L. Mordechai, M. Eisenberg, T. Newfield et al., "The Justinianic Plague: An Inconsequential pandemic?" *Proceedings of the National Academy of Sciences*, 116 (2019): 25546–25554.

> detection or using violence; but afterwards confusion and disorder everywhere became complete. For slaves remained destitute of masters, and men who in former times were very prosperous were deprived of the service of their domestics who were either sick or dead, and many houses became completely destitute of human inhabitants. For this reason, it came about that some of the notable men of the city because of the universal destitution remained unburied for many days [...]
>
> And when it came about that all the tombs which had existed previously were filled with the dead, then they dug up all the places about the city one after the other, laid the dead there, each one as he could, and departed; but later on those who were making these trenches, no longer able to keep up with the number of the dying, mounted the towers of the fortifications in Sycae, and tearing off the roofs threw the bodies there in complete disorder; and they piled them up just as each one happened to fall, and filled practically all the towers with corpses, and then covered them again with their roofs. As a result of this an evil stench pervaded the city and distressed the inhabitants still more, and especially whenever the wind blew fresh from that quarter.[123]

The catastrophe would almost certainly have disproportionately killed the lower classes, the poorer, as they would have had compromised immune systems due to malnutrition. As we saw earlier, during the Black Death, about 40 per cent of English people were killed, but it was about 80 per cent of the serfs and free labourers.[124]

People returned to simpler forms of life, based around mere

123 J. Mark, "Procopius on the Plague of Justinian: Text and Commentary," *World History Encyclopedia* (2020), https://www.worldhistory.org/article/1536/procopius-on-the-plague-of-justinian-text--comment/.

124 B. Dodds, "Patterns of decline: Arable production in England, France and Castile, 1370–1450," in B. Dodds and R. Britnell, (Eds.), *Agriculture and Rural Society After the Black Death: Common Themes and Regional Variations* (Hatfield: University of Hertfordshire Press, 2008).

survival in the here and now. They stopped passing on literacy and numeracy, they ceased to pass on knowledge of how to construct large buildings or roads, even animal husbandry, and how to produce larger livestock. All this was lost, as people became too survival-focused to bother with such notions or, with intelligence in decline, they could no longer comprehend them.[125] And the evidence is that Rome was suffering from the same declining intelligence that we are suffering from, as seen in the decline of per capita major innovations noted up until the collapse, and evidence from that time that the higher classes increasingly opted not to have children.[126] As this process continued, entire Roman cities were simply left to ruin. Seeds blew into cracks in the rock. Grass began to grow. It rotted, turning into earth. More grass grew and civilization—apart from a few refuges, such as Byzantium—collapsed. As this took place, more intelligent people—as indicated by their social class—fled, from England for example, to the areas still controlled by the Roman Empire, but these diminished until Byzantium was all that was left.[127]

The crucial point is that the population soared, and genetically weakened, under easy conditions, partly of its own creation, and then collapsed when the conditions changed or could not be sustained. This is what happens in the animal kingdom, and we are an advanced form of ape. So, wouldn't it be extraordinary if this wasn't what happened to us? The difference is that our population has become enormous, and enormously genetically weakened, due to the zoo that we have created for ourselves, so the collapse may well be correspondingly dramatic.

125 R. Lacey and D. Danziger, *The Year 1000: What Life Was Like at the Turn of the First Millennium* (London: Little Brown & Co, 1999).

126 See, Dutton and Woodley of Menie, *At Our Wits' End, op. cit.*

127 J. Moorhead, *The Roman Empire Divided: 400–700 AD* (London: Routledge, 2013).

About the Author

Edward Dutton is a researcher based in Oulu in northern Finland. Born in London in 1980, Dutton read Theology at Durham University, before completing a PhD in Religious Studies at Aberdeen University in 2006. He was made Docent of the Anthropology of Religion and Finnish Culture at Oulu University in 2011. In 2012, however, Dutton made the move to evolutionary psychology. He has been a guest researcher in the Psychology Department at Umeå University in Sweden and is research associate in scientific research at Tabuk University in Saudi Arabia. In 2020, he was made honorary Professor of Evolutionary Psychology at Asbiro University in Łódź, Poland. In 2023, Dutton was appointed Honorary Professor in the Institute of Psychology at the Russian Academy of Sciences. Dutton's research has been reported worldwide including in the *Daily Telegraph*, *The Sun*, *Le Monde*, and *Newsweek* and his books have been translated into various languages including Russian, German, Japanese and Turkish. His recent books include *The Past is a Future Country: The Coming Conservative Demographic Revolution* (with J.O.A. Rayner-Hilles, Imprint Academic, 2022) and *Sent Before Their Time: Genius, Charisma and Being Born Prematurely* (Manticore Press, 2022). Dutton can be found online at his award-winning channel "The Jolly Heretic."

Index

abortion 176–178, 185, 223–229, 247, 269–272, 282–283, 293
Addams, Jane 199
ADHD 8, 135, 140
adoption 17, 39, 45, 55, 112–115, 129, 211 234, 273
Africa 35, 55, 182–185, 215–217
agriculture 45, 60, 115–116
albinism 184–185
alleles 25–30, 80, 102, 116, 171, 265, 284–286
allergies 4, 6, 9–15, 170, 259
altruism 42–43, 75, 114, 122–123
anemia 29–30, 286
anxiety 107–111, 122, 134, 152, 158–162, 240, 256, 266, 271
Arkwright, Richard 49, 84
asthma 9–13, 152
atheism 120–121, 127, 133, 142, 161, 261
autism 32, 75, 112, 120–122, 127, 153–164, 239–240, 283–285

Balfour, Arthur 195, 210
Beveridge, William 196, 204
black people 190, 216, 225, 250, 257, 292
Buss, David 59
Byzantium 147, 264, 285, 299–300

cancer 9, 33, 168–170
Carl, Noah 243, 256
Catholicism 16, 215, 224, 277
Cattell, Raymond 243, 250
Charlton, Bruce 92–93, 106, 150
Chesterton, G. K. 204
China 61, 71, 88, 227, 247–248
Church, Major Archibald 201
Churchill, Winston 210
Clark, Gregory 68–71, 76, 79
Cochran, Gregory 44
conservatism 22, 108, 120, 128–132, 136, 140–142, 147, 160–162, 204, 218–221, 233, 255, 261, 266–268, 277, 280
Corbyn, Jeremy 258
Crick, Francis 209
criminals 39, 191, 194, 212, 240, 257, 274–275
Cummings, Dominic 257
cyclical history 39, 148, 174, 218, 262–266, 296

Darrow, Clarence 214
Darwin, Charles 37, 191, 202–203, 208
Darwin, Major Leonard 208
democracy 87, 205, 245, 290
DNA 25–27, 32–33, 62, 209, 278, 295

Douglass, Frederick 189–190
Down, John Langdon 34
Down's Syndrome 33–36, 175–179, 237–239, 272

dysgenics 15, 137, 172, 190, 202, 208, 227, 235, 242, 245, 248, 250, 253, 273, 290–292
dysphoria 143–152, 157–160, 164, 238, 263, 266, 273, 287

empathy 35, 108, 117, 122, 127–129, 138, 162, 239–240, 266, 271, 284
epilepsy 181, 197, 213, 221
epistasis 28, 55, 182, 274
ethnocentrism 42–45, 63, 114–119, 124, 144, 263, 292, 298
eugenicist 16, 86, 106, 188, 197–207, 215–216, 226, 236–237, 242–258, 293
euthanasia 214, 273–274
evolution 40–44, 85, 115, 122–124, 156, 202, 218, 263, 276, 287–288, 292

farmers 17, 61, 70
feminism 18, 86, 229, 263, 290
Fisher, Sir Ronald 47, 193–195, 208, 258
Flynn effect, the 89–90, 97
foundlings 73
Frazer, Sir James 217
Frost, Peter 74

Galton, Sir Francis 15, 191, 206
George, David Lloyd 203
Glad, John 61
gods 63, 121–124, 298
Grant, Madison 216
Greeks 186–190, 209, 265
Gregory, J. W. 216–217
group selection 39–45, 63, 66, 114, 117, 144, 148, 151, 167, 172, 262–263, 292

Haiselden, Harry J. 214
Haldane, J. B. S. 209
Hamilton, William 40, 286
Harpending, Henry 74
Harty, Martin 254
hemophilia 31
heritability 55, 72–75, 100, 112, 140, 146, 151, 168, 211, 243, 267
hierarchy 37–39, 45, 58, 64, 67–72, 130, 139, 159
Hinduism 63–66, 282, 295
homophobia 151
homosexuality 44, 151–161, 166, 263
Hsu, Steve 243
hunter-gatherers 56, 59–60, 115, 182–183
Huntington's Disease 25–30, 221
Huxley, Aldous 200
Huxley, Sir Julian 195

illegitimacy 67, 73, 205, 212
individualism 19, 127, 131, 142–143, 158, 161, 198, 233–242, 262–263, 282, 292, 295
infanticide 118, 182–184, 206, 247
infertility 25, 36, 62, 148–149, 173–176, 219, 276, 281
Inglehart, Ronald 139
IVF 275–279

Jaywick, Essex 21–22

Jensen, Arthur 91
Jews 61, 62, 177, 221–222, 244, 264
Johnson, Boris 257
Jordan, David Starr 208
Joseph, Sir Keith 253
Jussim, Lee 54

Keith, Sir Arthur 216
Keller, Helen 214
Kellogg, John Harvey 211
Keynes, John Maynard 195
Kong, Augustine 99
Kuttner, Robert 246

Lee Kuan Yew 227
liberalism 108, 128–134, 142, 158–159, 205, 233, 245, 261, 266
literacy 70, 76, 77, 78, 79, 80, 95, 101, 265, 300
Ludovici, Anthony 206
Lykken, David 251
Lynn, Richard 61, 247, 256, 277

MacGregor, Anthony James 246
Madison, Guy 75
maladaptivity 13, 25, 103, 136–137, 142–143, 162, 235, 263, 266, 276
marriage 18, 78, 118, 148–150, 187, 205, 216, 221, 225–229
Meisenberg, Gerhard 247, 257
Mermaid Syndrome 165
Merrick, Joseph 238
monogamy 67

Nazis 16, 106, 201, 206–207, 216, 221–223, 232, 242, 250, 255
Nyborg, Helmuth 146, 249

obesity 8, 22, 34, 152, 164, 168, 170–173
Olsson, Karl Johan 201

paraphilia 153, 156–157
pastoralists 59–60, 69, 115, 183
patriarchy 122, 144, 263, 298
Pearson, Adam 238
Pearson, Karl 193, 201, 202
Perkins, Adam 232, 244
Pinker, Steven 43
Pitt-Rivers, George 106, 195, 206–207
pleiotropy 24, 25, 45, 48, 182
polygamy 61, 67
Pound, John 69
Prader-Willi Syndrome 176
pregnancy 36, 56, 85–86, 112, 173, 222, 269–271, 275–276
Prince Edward, Duke of Kent 31
psychopathy 8, 74, 138–140, 225, 248, 270–271, 275, 293

racism 16, 134, 235, 238–239, 242–243, 250, 253, 293
Ramsay, Archibald 206
Ratcliffe, Nicky 3–5, 9–10, 165, 259
retardation, mental 29, 52, 176–178, 184, 197, 200, 210, 213, 221–223, 238, 254
Richmond, Mary 199
Robert Gayre of Gayre 246
Romans 78, 215, 263–264, 290, 297–300
Rutherford, Adam 65, 232, 293–295

Sabisky, Andrew 257
Saini, Angela 65, 295

schizophrenia 28, 122, 133–134, 138–140, 145, 152, 167, 221–223, 248, 290
segregation 165, 196, 199, 218, 220
Shockley, William 249–250
Siskind, Janet 58
Skrbina, David 286
slavery 61, 185–186, 189–190, 225, 299
socialism 198–201, 209, 222
socioeconomic status 22, 39, 55, 56, 59, 72, 87, 108, 117, 189, 195, 223, 227, 243, 260, 269, 270, 294
Spartans 45, 185–187
Spencer, Herbert 203
Stalin, Joseph 17
Starr, Jessica 126–127
stereotypes 54, 112, 243

taboos 19, 123–125, 205, 219, 232, 236, 242, 253
technology 63, 103–104, 149, 178, 196, 252, 259, 276–278
testosterone 38, 131, 152, 167, 173, 241
Thompson, James 254–257
Thomson, J. Arthur 18
tradition 205
transsexuality 152–156, 235, 292
Trump, Donald 125
trust, low 260, 291–293
Turner Syndrome 30–32

Unwin, J. D. 105

Vallejo-Nájera, Antonio 224

Weiss, Volkmar 69, 87
welfare 73, 87, 135, 212, 223, 227, 231, 244, 253, 261–267, 282
white people 215–217, 247, 250
Wilkinson, Ellen 201
Wilson, David Sloan 39
witchcraft 144, 184
wokeness 129, 134, 242–243, 250, 258, 267, 276–280, 292
WORDSUM 95–96

Yanomamö, the 56–57, 69, 114–115, 167, 183

Other Books by Imperium Press

NEEMA PARVINI	The Populist Delusion
JONATHAN BOWDEN	Why I Am Not a Liberal *(Studies in Reaction series)*
JOSH NEAL	American Extremist: The Psychology of Political Extremism
JOSEPH DE MAISTRE	Major Works, Volume I
FRIEDRICH LIST	The National System of Political Economy
GEORGE CARROWAY	Never Say D.I.E.: Three Dialogues on Diversity, Inclusion, and Equity
NICK LAND	The Dark Enlightenment *(Studies in Reaction series)*
	Folktales in the Indo-European Tradition *(Western Canon series)*
ROBERT FILMER	Patriarcha
	Poetic Edda *(Western Canon series)*
NUMA FUSTEL DENIS DE COULANGES	The Ancient City *(Traditionalist Histories series)*

Made in United States
North Haven, CT
10 November 2026

22088266R00183